Thermo-Mechanical Behaviour of Structural Lightweight Alloys

Thermo-Mechanical Behaviour of Structural Lightweight Alloys

Special Issue Editor
Guillermo Requena

MDPI • Basel • Beijing • Wuhan • Barcelona • Belgrade

Special Issue Editor
Guillermo Requena
German Aerospace Centre, Germany

Editorial Office
MDPI
St. Alban-Anlage 66
4052 Basel, Switzerland

This is a reprint of articles from the Special Issue published online in the open access journal *Materials* (ISSN 1996-1944) from 2018 to 2019 (available at: https://www.mdpi.com/journal/materials/special_issues/thermo_mechanical_lightweight_alloy)

For citation purposes, cite each article independently as indicated on the article page online and as indicated below:

LastName, A.A.; LastName, B.B.; LastName, C.C. Article Title. *Journal Name* **Year**, *Article Number*, Page Range.

ISBN 978-3-03921-387-0 (Pbk)
ISBN 978-3-03921-388-7 (PDF)

© 2019 by the authors. Articles in this book are Open Access and distributed under the Creative Commons Attribution (CC BY) license, which allows users to download, copy and build upon published articles, as long as the author and publisher are properly credited, which ensures maximum dissemination and a wider impact of our publications.

The book as a whole is distributed by MDPI under the terms and conditions of the Creative Commons license CC BY-NC-ND.

Contents

About the Special Issue Editor vii

Guillermo Requena
Special Issue: Thermo-Mechanical Behaviour of Structural Lightweight Alloys
Reprinted from: *Materials* **2019**, *12*, 2364, doi:10.3390/ma12152364 1

Cecilia Poletti, Romain Bureau, Peter Loidolt, Peter Simon, Stefan Mitsche and Mirjam Spuller
Microstructure Evolution in a 6082 Aluminium Alloy during Thermomechanical Treatment
Reprinted from: *Materials* **2018**, *11*, 1319, doi:10.3390/ma11081319 3

Philipp Wiechmann, Hannes Panwitt, Horst Heyer, Michael Reich, Manuela Sander and Olaf Kessler
Combined Calorimetry, Thermo-Mechanical Analysis and Tensile Test on Welded EN AW-6082 Joints
Reprinted from: *Materials* **2018**, *11*, 1396, doi:10.3390/ma11081396 18

Aleksander Kowalski, Wojciech Ozgowicz, Wojciech Jurczak, Adam Grajcar, Sonia Boczkal and Janusz Żelechowski
Microstructure, Mechanical Properties, and Corrosion Resistance of Thermomechanically Processed AlZn6Mg0.8Zr Alloy
Reprinted from: *Materials* **2018**, *11*, 570, doi:10.3390/ma11040570 40

Katrin Bugelnig, Holger Germann, Thomas Steffens, Federico Sket, Jérôme Adrien, Eric Maire, Elodie Boller and Guillermo Requena
Revealing the Effect of Local Connectivity of Rigid Phases during Deformation at High Temperature of Cast AlSi12Cu4Ni(2,3)Mg Alloys
Reprinted from: *Materials* **2018**, *11*, 1300, doi:10.3390/ma11081300 53

David Florián-Algarín, Raúl Marrero, Xiaochun Li, Hongseok Choi and Oscar Marcelo Suárez
Strengthening of Aluminum Wires Treated with A206/Alumina Nanocomposites
Reprinted from: *Materials* **2018**, *11*, 413, doi:10.3390/ma11030413 71

Serge Gavras, Ricardo H. Buzolin, Tungky Subroto, Andreas Stark and Domonkos Tolnai
The Effect of Zn Content on the Mechanical Properties of Mg-4Nd-xZn Alloys
(x = 0, 3, 5 and 8 wt.%)
Reprinted from: *Materials* **2018**, *11*, 1103, doi:10.3390/ma11071103 84

Zhengyuan Gao, Linsheng Hu, Jinfeng Li, Zhiguo An, Jun Li and Qiuyan Huang
Achieving High Strength and Good Ductility in As-Extruded Mg–Gd–Y–Zn Alloys by Ce Micro-Alloying
Reprinted from: *Materials* **2018**, *11*, 102, doi:10.3390/ma11010102 96

Gerardo Garces, Sandra Cabeza, Rafael Barea, Pablo Pérez and Paloma Adeva
Maintaining High Strength in Mg-LPSO Alloys with Low Yttrium Content Using Severe Plastic Deformation
Reprinted from: *Materials* **2018**, *11*, 733, doi:10.3390/ma11050733 108

About the Special Issue Editor

Guillermo Requena (Prof. Dr.) Head of the Department of Metals and Hybrid Structures of the German Aerospace Centre, and is Chair of "Metallic Structures and Materials Systems for Aerospace Engineering" at RWTH Aachen University. He has been working in the field of metallic structural materials for over 15 years. His research areas include:

- Microstructure—property relationships in structural light materials, e.g., Al, Mg, Ti, and TiAl alloys, MMC;

- Additive manufacturing of metals;

- 3D-imaging and diffraction techniques to investigate materials under manufacturing and service conditions.

Editorial

Special Issue: Thermo-Mechanical Behaviour of Structural Lightweight Alloys

Guillermo Requena [1,2]

1. Department of Metallic Structures and Hybrid Materials Systems, Institute of Materials Research, German Aerospace Centre, Linder Höhe, 51147 Cologne, Germany; Guillermo.Requena@dlr.de
2. Metallic Structures and Materials Systems for Aerospace Engineering, RWTH Aachen University, 52062 Aachen, Germany

Received: 17 July 2019; Accepted: 23 July 2019; Published: 25 July 2019

The need to reduce the ecological footprint of (water, land, air) vehicles in this era of climate change requires pushing the limits in the development of lightweight structures and materials. This requires a thorough understanding of their thermo-mechanical behaviour at several stages of the production chain. Moreover, during service, the response of lightweight alloys under the simultaneous influence of mechanical loads and temperature can determine the lifetime and performance of a multitude of structural components

The present Special Issue, formed by eight original research articles, is dedicated to disseminating current efforts around the globe aiming at advancing in the understanding of the thermo-mechanical behaviour of structural lightweight alloys under processing or service conditions. The two most prominent families of lightweight metals, namely aluminium and magnesium alloys, are represented with five and three contributions, respectively.

The work by Poletti et al. [1] deals with the evolution of the microstructure of an AA6082 alloy during thermo-mechanical processing. The production of wrought aluminium alloys usually comprises successive thermo-mechanical steps that involve complex physical phenomena at the microstructural level. Based on flow data and thorough microstructural observations they propose a physically-based constitutive model that can reproduce the behaviour of the alloys during cold and hot working over a wide range of strain rates. The same alloy was studied in [2] by Wiechmann et al. In this case, the authors studied the evolution of the microstructure and the mechanical behaviour of MIG welded joints applying several complementary ex situ and in situ experimental techniques. The results obtained in this work are a step forward to understand the influence of welding heat on the softening behaviour of this alloy. Also dealing with wrought Al alloys, although in a different alloy system, the work by Kowalski et al. [3] investigates the effect of low-temperature thermomechanical treatment (LTTT) on the microstructure, mechanical behaviour and corrosion resistance of a 7000 series AlZn6Mg alloy. Interestingly, they report conditions for LTTT which render better mechanical performance than conventional heat treatments. Moreover, they show that the electrochemical corrosion resistance of the alloy decreases with increasing plastic deformation, while, on the other hand, stress corrosion resistance is improved.

Bugelnig et al. [4] report on the effect of Ni concentration on the damage accumulation during high temperature tensile deformation of AlSi12Cu4Ni2–3 piston alloys. Using 3D and 4D synchrotron imaging the authors show that interconnecting branches within highly interconnected brittle networks of aluminides determine the damage evolution and ductility in these alloys. A load partition model that considers the loss of interconnecting branches within the rigid networks owing to damage is proposed to rationalize the experimental observations. The last contribution dealing with Al alloys addresses the characterization of A206 (AlCu4.5Mg) wires reinforced with 5 wt% of Al_2O_3 nanoparticles. This composite has potential application for TIG welding of aluminium [5]. Here, Florián-Algarín et al.

show that a significant strengthening is obtained by the addition of the so-called nanocomposite and that the addition of Al$_2$O$_3$ affected the electrical conductivity of the wires.

The contributions on Mg alloys are focused on development, processing and characterization of high strength alloys [6–8]. Gavras et al. [6] and Gao et al. [7] explore the improvement of mechanical performance by the addition of rare earth elements. Gavras et al. [6] investigated the evolution of strength and ductility at room and elevated temperature as a function of Nd addition to pure Mg and Mg–Zn alloys. They show that the binary MgNd4 alloy performs better than the ternary alloys up to an addition of 8 wt% of Zn. On the other hand, Gao et al. [7] address the effect of Ce addition on the microstructure of a MgGd7Y3.5Zn alloy. Ce promotes the formation of long period stacking order (LPSO) phases and show that an addition of 0.5 wt% Ce can result in an improvement of mechanical performance. Finally, Garcés et al. [8] present an in-depth report on the effect of severe plastic deformation of Mg-LPSO alloys. This group, which is in one of the pioneers in the study of LPSO-containing Mg alloys, shows that yield strengths similar to extruded conditions can be achieved with only half of the usual Y and Zn contents, owing to the grain refinement provoked by equal channel angular pressing.

I am confident that the readers will find the contributions to this special issue appealing since they address timely topics to further advance the development of structural Al and Mg alloys.

Acknowledgments: I personally would like to thank all contributors for the quality of their research articles as well as all reviewers for the time invested.

Conflicts of Interest: The authors declare no conflict of interest.

References

1. Poletti, C.; Bureau, R.; Loidolt, P.; Simon, P.; Mitsche, S.; Spuller, M. Microstructure Evolution in a 6082 Aluminium Alloy during Thermomechanical Treatment. *Materials* **2018**, *11*, 1319. [CrossRef] [PubMed]
2. Wiechmann, P.; Panwitt, H.; Heyer, H.; Reich, M.; Sander, M.; Kessler, O. Combined Calorimetry, Thermo-Mechanical Analysis and Tensile Test on Welded EN AW-6082 Joints. *Materials* **2018**, *11*, 1396. [CrossRef] [PubMed]
3. Kowalski, A.; Ozgowicz, W.; Jurczak, W.; Grajcar, A.; Boczkal, S.; Żelechowski, J. Microstructure, Mechanical Properties, and Corrosion Resistance of Thermomechanically Processed AlZn6Mg0.8Zr Alloy. *Materials* **2018**, *11*, 570. [CrossRef] [PubMed]
4. Bugelnig, K.; Germann, H.; Steffens, T.; Sket, F.; Adrien, J.; Maire, E.; Boller, E.; Requena, G. Revealing the Effect of Local Connectivity of Rigid Phases during Deformation at High Temperature of Cast AlSi12Cu4Ni(2,3)Mg Alloys. *Materials* **2018**, *11*, 1300. [CrossRef] [PubMed]
5. Florián-Algarín, D.; Marrero, R.; Li, X.; Choi, H.; Suárez, O.M. Strengthening of Aluminum Wires Treated with A206/Alumina Nanocomposites. *Materials* **2018**, *11*, 413. [CrossRef] [PubMed]
6. Gavras, S.; Buzolin, R.H.; Subroto, T.; Stark, A.; Tolnai, D. The Effect of Zn Content on the Mechanical Properties of Mg-4Nd-xZn Alloys (x = 0, 3, 5 and 8 wt.%). *Materials* **2018**, *11*, 1103. [CrossRef] [PubMed]
7. Gao, Z.; Hu, L.; Li, J.; An, Z.; Li, J.; Huang, Q. Achieving High Strength and Good Ductility in As-Extruded Mg–Gd–Y–Zn Alloys by Ce Micro-Alloying. *Materials* **2018**, *11*, 102. [CrossRef] [PubMed]
8. Garces, G.; Cabeza, S.; Barea, R.; Pérez, P.; Adeva, P. Maintaining High Strength in Mg-LPSO Alloys with Low Yttrium Content Using Severe Plastic Deformation. *Materials* **2018**, *11*, 733. [CrossRef] [PubMed]

© 2019 by the author. Licensee MDPI, Basel, Switzerland. This article is an open access article distributed under the terms and conditions of the Creative Commons Attribution (CC BY) license (http://creativecommons.org/licenses/by/4.0/).

Article

Microstructure Evolution in a 6082 Aluminium Alloy during Thermomechanical Treatment

Cecilia Poletti [1,*], Romain Bureau [2], Peter Loidolt [3], Peter Simon [4], Stefan Mitsche [5] and Mirjam Spuller [6]

1. Institute for Materials Science, Joining and Forming, Graz University of Technology, Kopernikusgasse 24, 8010 Graz, Austria
2. Advanced Materials and Mechanical Testing, French-German Research Institute of Saint-Louis, 5 rue du Général Cassagnou, 68300 Saint-Louis, France; romain.bureau@isl.eumailto
3. Institute for Process and Particle Engineering, Graz University of Technology, Inffeldgasse 13/3, 8010 Graz, Austria; peter.loidolt@tugraz.at
4. AMAG Austria Metall AG, Lamprechtshausenerstrasse 61, P.O. Box 3, 5282 Braunau-Ranshofen, Austria; peter.simon@amag.at
5. Institute of Electron Microscopy and Nanoanalysis of the TU Graz (FELMI), Graz Centre for Electron Microscopy (ZFE Graz), Steyrergasse 17, 8010 Graz, Austria; smitsche@tugraz.at
6. Erich Schmid Institute of Materials Science of the Austrian Academy of Sciences, Jahnstrasse 12, 8700 Leoben, Austria; mirjam.spuller@oeaw.ac.at
* Correspondence: cecilia.poletti@tugraz.at; Tel.: +43-316-873-1676

Received: 31 May 2018; Accepted: 26 July 2018; Published: 30 July 2018

Abstract: Thermomechanical treatments of age-hardenable wrought aluminium alloys provoke microstructural changes that involve the movement, arrangement, and annihilation of dislocations, the movement of boundaries, and the formation or dissolution of phases. Cold and hot compression tests are carried out using a Gleeble® 3800 machine to produce flow data as well as deformed samples for metallography. Electron backscattered diffraction and light optical microscopy were used to characterise the microstructure after plastic deformation and heat treatments. Models based on dislocation densities are developed to describe strain hardening, dynamic recovery, and static recrystallisation. The models can describe both the flow and the microstructure evolutions at deformations from room temperatures to 450 °C. The static recrystallisation and static recovery phenomena are modelled as a continuation of the deformation model. The recrystallisation model accounts also for the effect of the intermetallic particles in the movements of boundaries.

Keywords: thermomechanical treatment; aluminium alloy; recovery; recrystallisation; dislocations; materials modelling

1. Introduction

The production process of 6xxx series aluminium sheets consists in a succession of thermomechanical steps designed to improve the strength of the product while reaching the desired geometry. The initial billet with its specific chemical composition is produced by continuous or batch casting. Each subsequent step brings irreversible changes in the microstructure that directly affect the mechanical properties of the material. A combination of recrystallised, finely grained, and precipitation hardened microstructure brings the best mechanical strength to the sheet while preserving a reasonable ductility for further shaping processes.

It is now a well-established practice to model the industrial processes with finite element methods, which require material data as an input. Modelling allows to roughly calculate the properties of the final product, supporting the design and optimisation of the production processes. As the models can only be as good as our understanding of the physical phenomena they are meant to represent, an experimental investigation is always needed to validate their output and to understand the underlying physical phenomena.

Phenomenological models consist in setting up a constitutive equation linking the flow stress to the strain, the strain rate, and the temperature, and optimising it so that it best represents the main features of the flow curves. The equation usually features a power law dependency for the strain and the strain rate, and an activation energy for the thermal dependency [1–4]. Such models present the advantages of being easy to set up and requiring almost no computational power, but they do not provide any insight on the physics of the problem at hand.

Physical models are more deeply connected to the microstructure evolution. The modification of the microstructure depends essentially on the material itself, its initial state, and its thermomechanical history. In the last decades, metallurgists have developed models to describe the strain hardening of metallic alloys during forming processes using physical-based microstructural approaches. Such approaches usually consist of three main features [5–8]; a set of independent internal variables representative of the microstructure (classically dislocation densities, subgrain sizes, precipitation state, etc.), the evolution rates of these variables, and a constitutive equation to link the microscopic variables with the flow stress of the material. The difficulty in observing the underlying mechanisms responsible for the variable evolution often leads to the appearance of a large amount of model parameters [7,8].

During deformation, the microstructure of metallic materials develops permanently, leading to important dynamic variations in the macroscopic stress required to further deform the material. Strain hardening, for example, results directly from the multiplication of microscopic defects such as dislocations. These variations essentially depend on the strain, the strain rate, and the temperature of work. After deformation, for example in between passes in a rolling process, the microstructure may undergo static recovery, that is, the annihilation and rearrangement of microstructural defects that also affects the flow stress of the material. Flow stresses developed during hot rolling at elevated temperatures can be modelled using the total dislocation density as a single internal variable to represent the microstructure [6], but modelling the behaviour of aluminium alloys from room to moderate temperatures requires at least two kinds of internal variables [6]. In rolled aluminium products, static recrystallisation occurs after cold rolling during a recrystallisation or a solution treatment. The dominant mechanism for the nucleation new grains is strain-induced boundary migration [9,10], whereby subgrains lying on the boundaries of existing deformed grains bulge into the neighbouring grain and grow further after reaching the critical size for nucleation. As the motion of boundaries is a diffusional process, it is influenced by the temperature. The strain grade directly influences the number of potential nuclei being available for further growth.

In this work we propose a physically based constitutive model applicable to cold and hot working over a wide range of strain rates using three internal variables. Two kinds of dislocation densities that can evolve are responsible for the strain hardening, while the third kind accounts for the deformation and can be derived directly from the Orowan equation [11]. Simple evolution rates are split in a dynamic part and a static part, enabling static recovery. A subsequent recrystallisation model featuring the same internal variables was developed that considers the level of strain reached during earlier deformation and encompasses the competition between recrystallisation and static recovery. The physical phenomena were derived and validated from experimental results obtained during cold and hot deformation and further recrystallisation of a 6082 aluminium alloy.

2. Materials and Methods

2.1. Material

A commercial 6082 aluminium alloy with the chemical composition shown in Table 1 was studied. The material was delivered after hot rolling into a plate of thickness 3.9 mm and subsequent air cooling at room temperature. No homogenisation treatment was applied before rolling.

Table 1. Chemical composition of the commercial AA6082, in weight percent.

Si	Fe	Cu	Mn	Mg	Cr	Ni	Zn	Ti	Al
0.88	0.39	0.07	0.43	0.81	0.02	0.01	0.04	0.04	Balance.

2.2. Experimental Methods

Samples of size 10 mm in the rolling direction and 20 mm in the transverse direction were cut out of the plate and compressed down to a thickness of 1.5 mm in plane strain condition using a Gleeble® 3800 machine (Dynamic Systems Inc., 323 NY 355 Poestenkill, NY, USA). The experiments were carried out between 25 °C and 400 °C, at strain rates of 0.01, 0.1, 1, and 10 s^{-1}. The temperature was controlled by a J type thermocouple welded on the surface of the samples. Samples deformed at room temperature and 10 s^{-1} were then annealed in an oven at 300 °C and 400 °C for 10 s, 1 min, 5 min, and 1 h to induce recrystallisation.

Light optical microscopy was carried out to determine the grain shape and size. Polarised light was used after metallographic preparation of the sample with Barker's reagent (5 mL HBF4 48%vol in 200 mL water). Additionally, the precipitates were characterised in scanning electron microscope (SEM) Zeiss Ultra 55 (Carl-Zeiss AG, Oberkochen, Germany) with the use of a backscattered electron detector (BSE) which shows material contrast, and an energy dispersive X-ray detector (EDS) EDAX Genesis (EDAX Business Unit AMETEK GmbH, Weiterstadt, Germany). For these characterisations, the SEM was operated at a primary beam energy of only 6 kV to gain high surface sensitive measurements. The intermetallic phases were detected using BSE detector. The density N_S of particles intersecting the sample surface, as well as the average area of intersection $\overline{A_P}$ of individual particles, were readily measured from the obtained micrographs. Assuming many randomly distributed spheres of radius R_P, the averaging of the area of intersection of individual particles with the sample surface over $[-R_P; R_P]$, is written as:

$$\overline{A_P} = \frac{2\pi}{3} R_P^2 \tag{1}$$

For a homogeneous distribution of particles, the volume fraction of particles F_V reads:

$$F_V = \frac{2\pi}{3} R_P^2 N_S = \overline{A_P} N_S \tag{2}$$

The as-received, deformed, and annealed samples were investigated by electron backscattered diffraction (EBSD) coupled with EDS. These investigations provided information about the grain and subgrain structure and of the intermetallic phases. All coupled EBSD–EDS investigations were performed at a primary beam energy of 20 kV on the Zeiss Ultra equipped with an EBSD system from EDAX-TSL. A tolerance angle of 11° was used to determine the subgrain structure.

Finite element simulations of the plane strain tests were carried out with the software DEFORM™ 2D (Scientific Forming Technologies Corporation, Columbus, OH, USA) to obtain the distribution of the equivalent strain values. The recrystallisation grade and the recrystallised grain size were characterised in the regions of strain 1 and 1.5 for both annealing temperatures. Grain orientation spread (GOS) maps were produced from EBSD measurements to distinguish recrystallised from non-recrystallised grains [12] defining a maximum spread limit of 3°.

2.3. Modelling Methods

2.3.1. Microstructure Representation

The microstructure is assumed to be composed of well-defined subgrains, whose walls and interiors are populated with dislocations of respective densities ρ_w and ρ_i. It is emphasised that, although ρ_w usually stands for the dislocation density within the subgrains [7,8], it represents here the total length of wall dislocation per unit volume of the material. The later definition yields lower densities than the former. Additionally, mobile dislocations of density ρ_m travel across several subgrains before being stored in some manner, accounting for the macroscopic strain. The total density of dislocations ρ_t hence reads:

$$\rho_t = \rho_w + \rho_i + \rho_m \quad (3)$$

The subgrain size δ in the deformed material can be calculated out of the dislocation densities. Two methods are available in the literature [9]. If the subgrain boundaries are assumed to be tilt boundaries and if the wall dislocation density is averaged over the microstructure, then:

$$\delta = \frac{\kappa \bar{\theta}}{b \rho_w} \quad (4)$$

κ is a shape factor and $\bar{\theta}$ is the average crystal orientation difference between each side of the boundary. This approach works fine at low to intermediate temperatures.

2.3.2. Constitutive Equation

The governing constitutive equation is chosen to have the following form:

$$\sigma = M \left(\tau_{ath} + \tau_{eff} + \tau_d \right) \quad (5)$$

where σ is the flow stress of the material and M is the Taylor factor, accounting for the polycrystalline nature of the material [13]. The athermal shear stress τ_{ath} resolved on the slip plane translates the long-range interaction of dislocations via their elastic strain field, and reads [14]:

$$\tau_{ath} = \alpha \mu b \sqrt{\rho_t} \quad (6)$$

α is a stress constant, μ is the temperature dependent shear modulus, and b is the Burgers vector. The effective resolved shear stress τ_{eff} is the additional stress required for mobile dislocations to be able to cut through the forest of dislocations cutting the slip plane and hindering them locally on their way through the microstructure. It is given by:

$$\tau_{eff} = \frac{Q}{V} + \frac{k_B T}{V} \exp\left(\frac{v_m}{L \nu_D} \right). \quad (7)$$

L is the mean free path of dislocations, ν_D is the Debye vibrational frequency of the material, Q is the energy barrier of forest dislocations, V has the dimension of a volume and is classically referred to as the activation volume, k_B is the Boltzmann constant, and T is the temperature. The glide velocity v_m of mobile dislocations is given by the Orowan equation:

$$M \dot{\epsilon} = \rho_m b v_m \quad (8)$$

where $\dot{\epsilon}$ is the rate of strain. The contribution of the intermetallic phases to the resolved shear stress τ_d is the Orowan stress [15]:

$$\tau_d = \frac{\mu b \sqrt{V_d}}{R_d} \quad (9)$$

V_d and R_d, respectively, being the volume fraction of dispersoids and their equivalent radius.

2.3.3. Rate Equations

The evolution rates of ρ_i and ρ_w are each given by a Kocks–Mecking type of equation [16] supplemented by a static annihilation term [17]:

$$\frac{\partial \rho_x}{\partial t} = \dot{\epsilon}\left(\frac{h_{1,x}}{b}\sqrt{\rho_x} - h_{2,x}\rho_x\right) - h_{3,x} D\left(\rho_x - \rho_{x,eq}\right)^2 \qquad (10)$$

with $x = i, w$. $h_{1,x}, h_{2,x}$ and $h_{3,x}$ are dimensionless model parameters. $\rho_{x,eq}$ is an equilibrium dislocation density of a fully recrystallised material. D is the diffusion coefficient:

$$D = b^2 \nu_D \exp\left(-\frac{Q_{self}}{k_B T}\right) \qquad (11)$$

where Q_{self} is the activation energy for self-diffusion. In Equation (10), the evolution rates are split into a dynamic part, linked to the strain rate, and a static part, diffusion driven. Although the kinetics of static mechanisms are negligible with respect to dynamic mechanisms, such a form of the model allows for diffusional phenomena when the strain rate is low or null and the temperature high enough.

2.3.4. Recrystallisation Model

The driving force for static recrystallisation and static recovery being related to the local stored energy, both processes happen simultaneously and competitively during annealing. The dislocation density decrease due to static recovery can be calculated by setting $\dot{\epsilon} = 0$ in Equation (10). If the subgrain growth is assumed to be driven by capillarity [9]:

$$\frac{\partial \delta}{\partial t} = M_s \frac{1.5\gamma_s}{\delta}. \qquad (12)$$

M_s being the mobility of low angle grain boundaries and γ_s their specific energy, given by a Read and Shockley relationship [18] of the form:

$$\gamma_s = \frac{\mu b \theta}{4\pi(1-\nu)} \ln\left(\frac{e\theta_c}{\theta}\right) \qquad (13)$$

where θ_c is the critical orientation difference for a low angle boundary to turn into a high angle boundary, e is the natural exponential, and ν is the Poisson coefficient of the material. Since the mobile dislocation density is negligible with respect to the subgrain interior dislocation density, the difference in stored volume energy ΔE between non-recrystallised and recrystallised grains reads:

$$\Delta E = 0.5 \mu b^2 \rho_i + \frac{1.5\gamma_s}{\delta} \qquad (14)$$

Capillarity also tends to promote the growth of recrystallised grains, which translates in a driving pressure P_C:

$$P_C = \frac{1.5\gamma_g}{D} \qquad (15)$$

where γ_g is the specific energy of high angle grain boundaries and D is the mean grain size. Second phase particles hinder the movement of boundaries by exerting a retarding pressure P_Z on them. The Zener mechanism [19] yields the following equation:

$$P_Z = \frac{3\gamma_g V_d}{2R_d} \qquad (16)$$

The total driving pressure for recrystallisation P is then classically given by:

$$P = \Delta E + P_C - P_Z \qquad (17)$$

The growth rate of recrystallised grains $\dot{G}$ can be written as:

$$\dot{G} = M_g P \qquad (18)$$

M_g being the mobility of high angle grain boundaries. The grain radius is then calculated by integrating Equation (18) over time.

$$D = 2 \int_{t_0}^{t} \dot{G}(t') dt' \tag{19}$$

The factor 2 accounts for the fact that D is the diameter of recrystallised grains. The incubation time t_0, that is, the time it takes for the nuclei to reach the critical size and start growing, is temperature dependent and must be adjusted. According to the JMAK theory [20], the extended recrystallised volume can be calculated as the product of the number N of nuclei available in the microstructure with the volume of a recrystallised grain. For site saturated nucleation, meaning that the N nuclei are created during deformation prior to annealing, this yields an equation of the type:

$$V_{ext} = NfD^n \tag{20}$$

where f is a shape factor and n the Avrami exponent. The fraction of recrystallised material X is then given by:

$$X = 1 - \exp(-V_{ext}) \tag{21}$$

The number of nuclei reaching the critical size for nucleation is written as follows:

$$N = N_0 \epsilon \exp\left(-\frac{\gamma_g b^2}{k_B T}\right) \tag{22}$$

where N_0 is a model parameter. The activation energy is taken as the product of the specific energy of high angle boundaries with a typical area, as for strain induced boundary migration to happen, subgrains have to bulge into the neighbouring grain and displace the existing boundary. Grain growth after recrystallisation is driven only by capillarity. Hence, the driving pressure is the same as in Equation (17), without the term ΔE. The grain size is again given by integrating the growth rate, but now starting at the time of end of recrystallisation.

2.3.5. Parameter Initialisation

The parameters of the flow stress model were initialised as follows: $\alpha = 0.5$, $b = 0.286$ nm [21], $M = 3.06$ [22], $\nu_D = 1.5 \times 10^{13}$ s^{-1}, $\nu = 0.33$, and $Q_{self} = 0.98$ eV. The shear modulus reads $\mu = (84.8 - 4.06 \times 10^{-2} T)/(2(1+\nu))$ in GPa (temperature in K) [23]. The microstructure being initially fully recrystallised, ρ_i and ρ_w were initially taken equally low and equal to 10^{10} m/m^3. Since the model must be able to work out the yield stress of the material, a least square optimisation method was run on ρ_m, V_{act}, and Q to best capture the temperature and strain rate dependency of the yield stress, with ρ_m let free to vary with the strain rate. The following values were worked out: $\frac{\rho_m}{\rho_m^*} = e^{2.28} \left(\frac{\dot{\epsilon}}{\dot{\epsilon}^*}\right)^{0.65}$ 1/m^2, $V_{act} = 1 - 50 \times 10^{-27}$ m^3, and $Q = 1.44$ eV. The rate parameters $h_{1,x}$ and $h_{2,x}$ ($x = w, i$) were optimised to best capture the strain hardening of the material, while the $h_{3,x}$ ($x = i, w$) were taken equal to 1.

The parameters of the recrystallisation model were initialised as follows: $\gamma_g = 0.324$ J·m^{-2} [9]. It was assumed that the subgrain misorientation already reaches, at very small strain values, a mean value of $\theta = 3°$ [8]. The critical misorientation was selected as $\theta_c = 15°$. M_g classically has an Arrhenius expression of the form $M_g = M_0 \exp(-Q_g/k\_BT)$ and the literature provides wide ranges of values for both M_0 and Q_g. The following was determined from our recrystallisation experiments and used here: $M_0 = 3.1 \times 10^{-9}$ m^4·J^{-1}·s^{-1} and $Q_g = 0.50$ eV. The mobility of the low angle grain boundaries M_s was taken equal to 0.02 M_g [24]. The parameter N_0 was equal to 7×10^{10}.

3. Results

3.1. Microstructural Features

3.1.1. As-Received Material

Figure 1 shows the microstructure of the AA6082 material in the as-received condition, that is, hot rolled and air cooled. The microstructure consists of disc-like grains partially recrystallised.

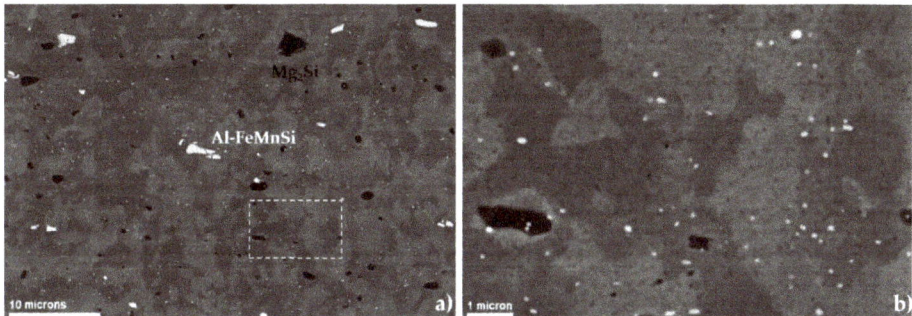

Figure 1. Microstructure of the as-received material observed by means of scanning electron microscopy (SEM) in backscattered electron mode (BSE) showing Al-FeMnSi (large and small) in white, and Mg$_2$Si in black. (**a**) Overview and (**b**) detail of rectangle shown in (**a**).

Three types of intermetallic phases were identified: 1 vol % of stable β-phase (Mg$_2$Si) with a mean radius of 0.35 μm, a population of large Al-FeMnSi phases of 1.7 vol % with a mean radius of 2.5 μm, and 0.53 vol % of Al-FeMnSi phases with a mean radius of 60 nm. Only this last population of Al-FeMnSi was considered to produce strengthening by Orowan mechanism and therefore, V_d = 0.53 vol % and R_d = 60 nm were used in Equations (9) and (16). The other particles could not produce strengthening due to the low amount and large size. It is assumed that large Mg$_2$Si existed only in the stable form due to the slow cooling after rolling.

3.1.2. Plastically Deformed

Figure 2 shows the microstructure, the strain, and the hardness distribution within a plane strain sample after cold deformation. Elongated grains and substructure formation can already be detected with light optical microscopy. The deformation under non-frictionless conditions provoked the appearance of a double cross of strain concentration.

The finite element calculations (Figure 3), show that the nominal strain of 1 is achieved in between the deformation crosses, whereas a local strain of 1.5 is achieved in the middle of each cross.

The substructure developed after plane strain deformation at room temperature, as well as the location of the intermetallic phases, are shown in Figure 4. The formed cells have a mean diameter of 1.5 μm and are surrounded by high angle and low angle grain boundaries. Cell size is heterogeneous within a grain and among grains.

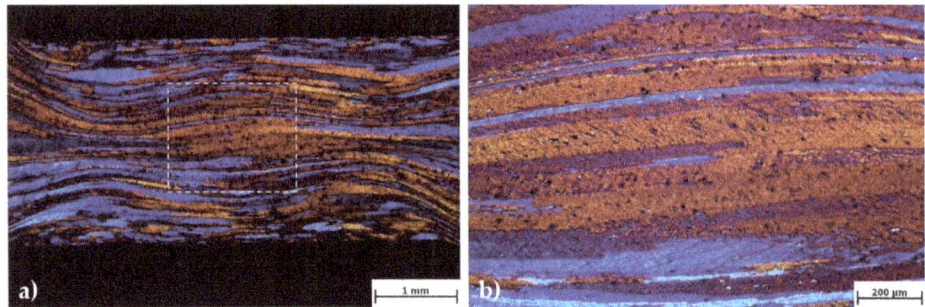

Figure 2. Elongated grains (**a**) and deformation bands (**b**) observed in the plane strain samples after cold deformation (optical microscopy). Nominal strain = 0.7.

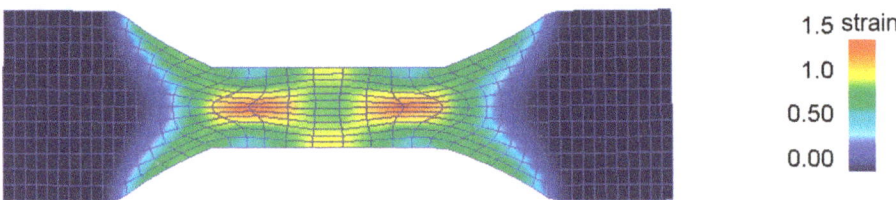

Figure 3. Finite element simulation using DEFORM™ 2D showing the strain distribution within the plane strain sample after cold deformation at a nominal strain of 1.

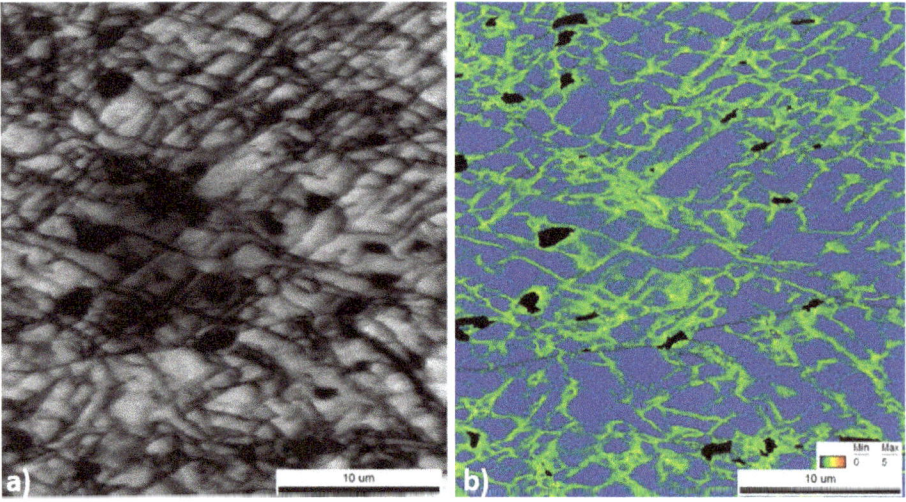

Figure 4. Microstructure of AA6082 after cold deformation determined by electron backscattered diffraction (EBSD) showing the cells in the (**a**) image quality (IQ) as well as (**b**) in the Kernel map. Black non-indexed areas represent the intermetallic positions.

The evolution of the dislocation densities and the subgrain size at room temperature are shown in Figure 5. The steady hardening occurring during cold deformation is produced by the increment of the total dislocation density. While the dislocation densities in the subgrain interiors saturate at small strains, the wall dislocation density keeps increasing. The latter can be interpreted as a continuous decrease in the subgrain size, since the model assumes saturation of the misorientation between subgrains at 3°. The experimental data from Gil Sevillano et al. [25], as well as the own measured data point are in good agreement with the modelled data.

The modelled cell interior and cell wall dislocation are shown in Figure 6. In general, it can be observed that ρ_i saturates very rapidly, especially by increasing the temperature, although with very low influence of the strain rate in the levels of the calculated values and in the investigated range of strain rates. The wall dislocation density saturates at low strain rates and high temperatures due to a large dynamic recovery, while it keeps increasing when the strain hardening dominates.

In agreement with the evolution of the wall dislocation densities, the subgrain size (Figure 7) reaches a plateau at high temperatures and low strain rates, when the subgrains are predicted to be the largest. The results agree with subgrain size of a similar AA6082 material determined experimentally by EBSD after hot compression tests [26].

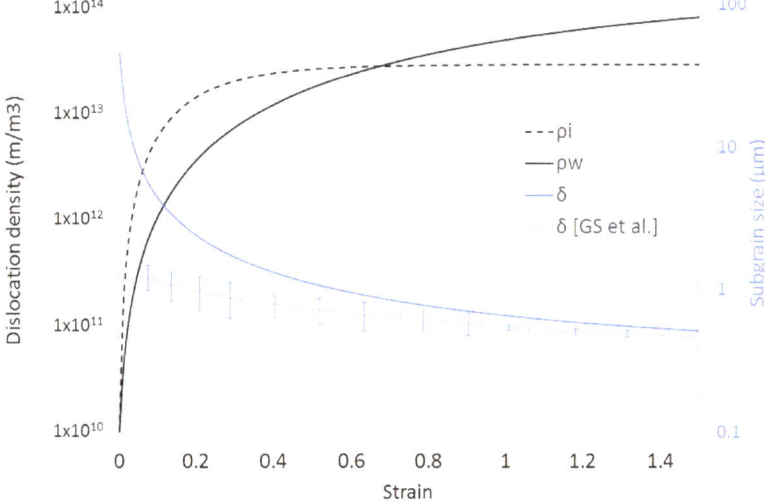

Figure 5. Modelled dislocation densities and subgrain size developed during deformation at room temperature and 10 s^{-1} of strain rate. Experimental data from Reference [25] show good correlation with the literature.

3.1.3. Recrystallisation

The recrystallisation behaviour of the material was studied after cold deformation in the regions of strain 1 and 1.5 for annealing at 300 °C and 400 °C. Unique grain colour maps obtained from EBSD data are shown in Figure 8. Recrystallisation was not observed in the samples treated at 300 °C before 20 min, and even then, only in the region of the highest strain. The microstructure is completely recrystallised after 1 h of annealing. The grains are smaller in the regions of larger strain, where the number of nuclei of recrystallised grains is higher. Elongated recrystallised grains were observed in the region of lower strain. The grain boundary migration is stopped by the Fe/Mn rich aluminides aligned in the direction of deformation. The microstructure is completely recrystallised after 5 min of annealing in the samples annealed at 400 °C, and no further grain growth is observed. The mean grain size after recrystallisation is larger at 300 °C than at 400 °C, which can be explained as follows: the critical size for nucleation is reached sooner at 400 °C than at 300 °C, leading to a larger number of nuclei. Their boundaries impinge rapidly upon each other, preventing further grain growth.

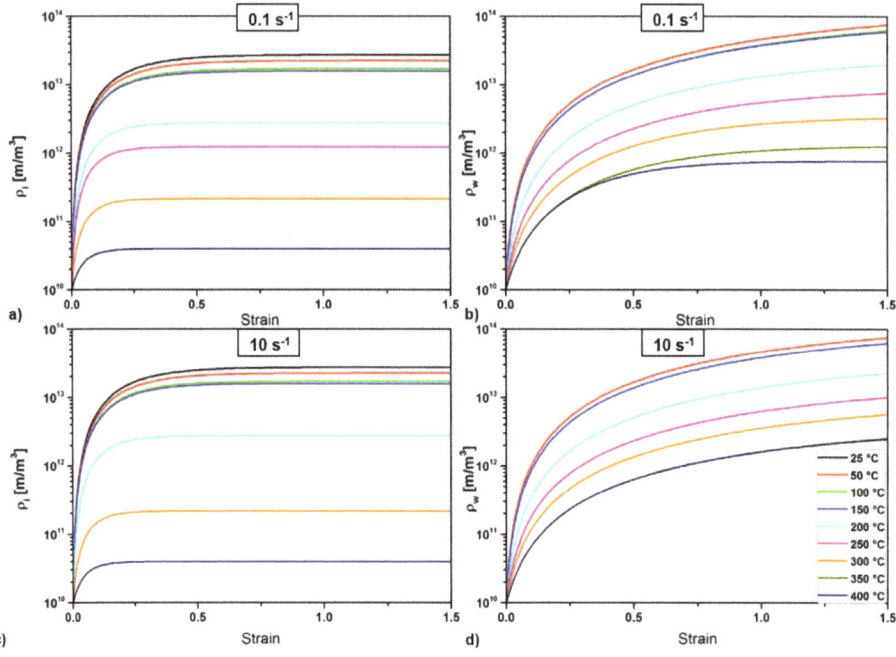

Figure 6. Evolution of the dislocation densities (modelled) as a function of temperature and for two selected strain rates. (**a,b**) interiors and walls dislocation densities, respectively, during deformation at 0.1 s^{-1}, and (**c,d**) interiors and walls dislocation densities, respectively, during deformation at s^{-1}.

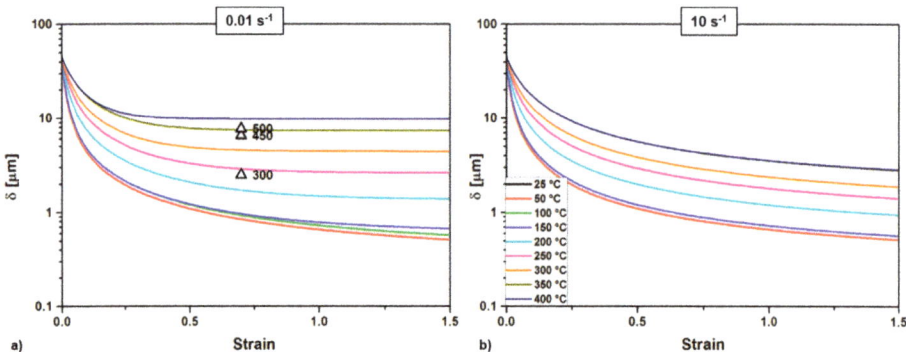

Figure 7. Calculated subgrain size evolution as a function of temperature and for two strain rates: 0.01 (**a**) and 10 s^{-1} (**b**). Comparison with experimental data from Reference [26].

The evolution of the recrystallised fraction during annealing at 300 °C and 400 °C is given in Figure 9. The effect of the cold deformation is reproduced by the model: a larger strain results in a faster recrystallisation and smaller grains. A good agreement is found between the model and the experimental results. The grain size reaches a constant value due to a complete Zener pinning of the high angle grain boundaries by the larger Al-FeMnSi phases. The plateau is reached earlier when the annealing temperature is 400 °C.

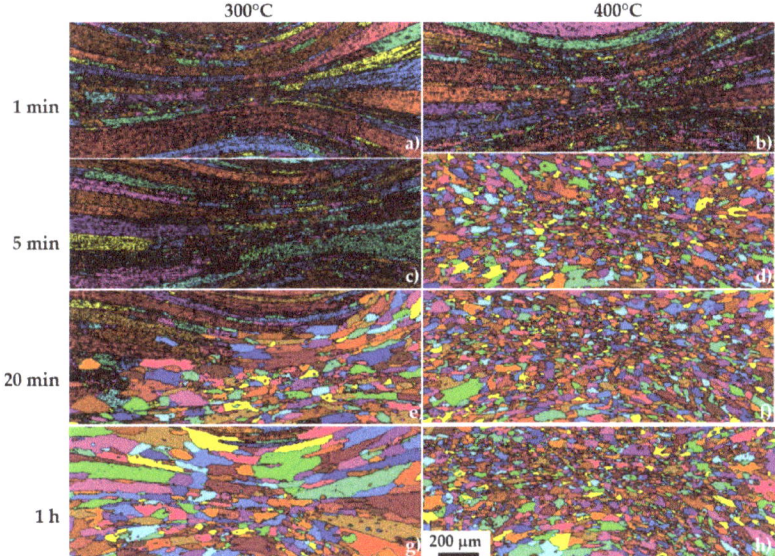

Figure 8. Unique grain colour map of the samples annealed at 300 °C (**a,c,e,g**), and 400 °C (**b,d,f,h**), for 1, 5, 20, and 60 min, respectively, obtained after EBSD measurements. The scale bar is applied to all sub-figures.

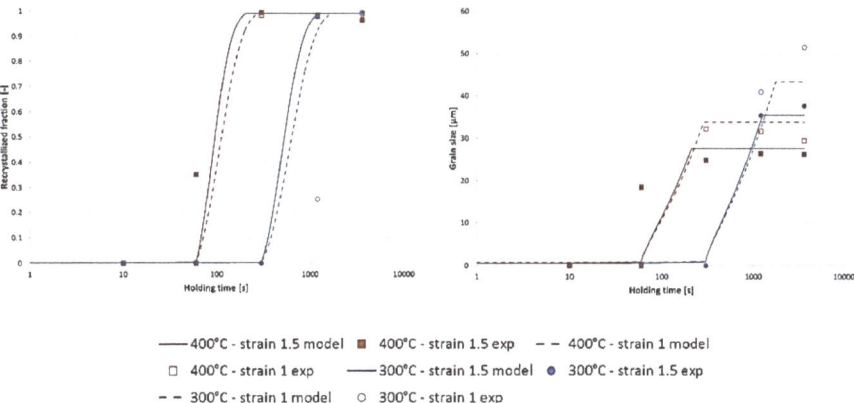

Figure 9. Recrystallised fraction (**left**) and grain size (**right**) during annealing following cold deformation at two different strains.

3.2. Flow Stresses

Experimental and modelled flow stresses are drawn in Figure 10. The modelled flow stresses show a good agreement with the experimental ones. A typical behaviour for aluminium alloys is observed: larger stresses at lower temperatures and higher strain rates, large strain hardening at low temperatures, and an increasing strain rate sensitivity by increasing the temperature. Although the model catches the temperature and strain rate effect on the stresses, in general the steady state is reached before than in the experimental data.

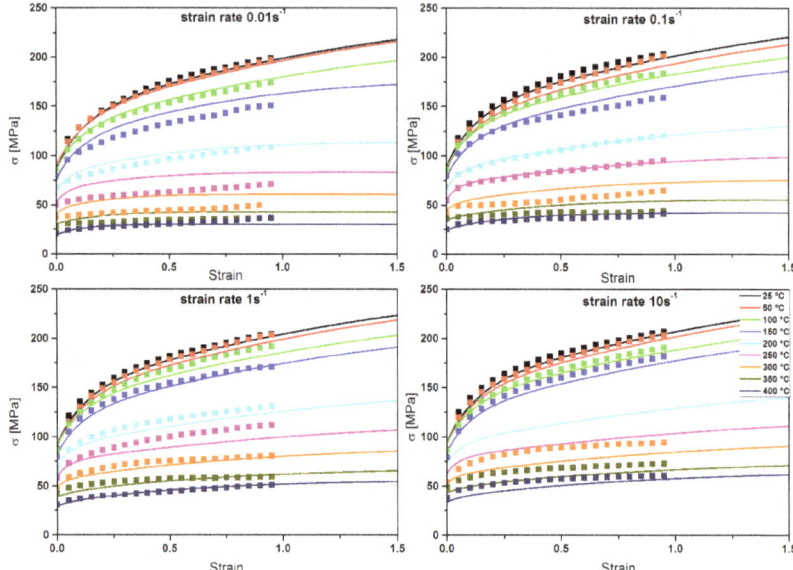

Figure 10. Measured (squares) and modelled (lines) flow stress at different strain rates and for temperatures ranging from 25 °C to 400 °C.

The flow stresses were obtained as the sum of different contributions as given by Equation (5). The effective and the athermal stresses are reported in Figure 11a,c and Figure 11b,d, respectively. The strain rate has no influence on them under 150 °C since the flow stress has no dependency on the strain rate at low deformation temperatures.

The athermal stress grows rapidly below 200 °C, and more slowly when the temperature increases. Saturation of the stress is observed at low strain rates and moderate to high temperatures. This is because the static part of the evolution rates is not negligible under these deformation conditions. The effective stress saturates at all temperatures and strain rates. The temperature has no influence below 100 °C. Above 100 °C the curves are regularly spaced, following the linearity of Equation (7). The contribution of the intermetallic phases to the resolved shear stress τ_d are assumed to be constant for each temperature since the volume and size of precipitates considered for its calculation (small Mn and Fe rich aluminides) do not change. The values of τ_d decrease linearly with the temperature from 8.2 MPa at 25 °C to 6.5 MPa at 400 °C.

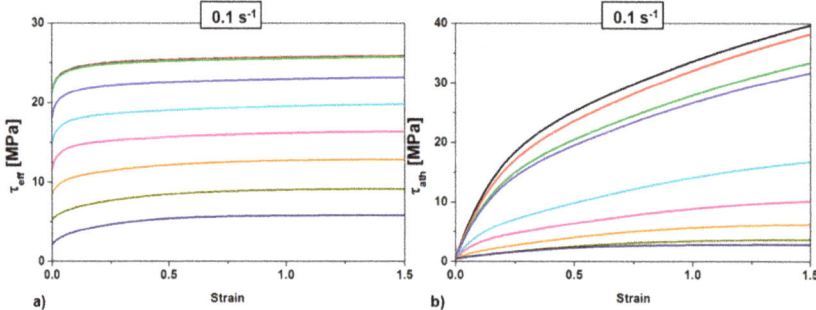

Figure 11. *Cont.*

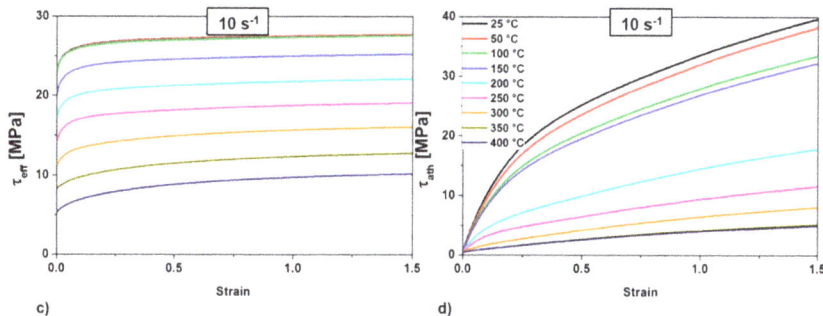

Figure 11. Effective and athermal stresses calculated for all temperatures and (**a**,**b**) 0.1 s^{-1} and (**c**,**d**) 10 s^{-1}.

4. Discussion and Conclusions

The flow stress model uses a relatively common approach to constitutive modelling. A strong hypothesis, based on the Orowan equation, is that the mobile dislocation density depends only on the strain rate, and remains constant during deformation at constant strain rate. The dislocation rate parameters are plotted against the temperature in Figure 12. The decrease of the hardening parameter $h_{1,i}$ with the temperature indicates that the storage of dislocations becomes less effective when the temperature increases, that is, that the mobile dislocations travel a longer distance before being immobilised, indicating an improved capacity for bypassing local obstacles when the temperature increases. The results also indicate that the hardening due to subgrain formation is promoted at lower temperatures since $h_{1,w}$ decreases with increasing temperatures. As expected, $h_{2,i}$ increases with the temperature, meaning that the dynamic recovery of cell interior dislocations is promoted. Interestingly, $h_{2,w}$ does not depend on the temperature. This seems to indicate that the wall dislocations, being already arranged in a configuration of low energy, are not appreciably affected by dynamic climb.

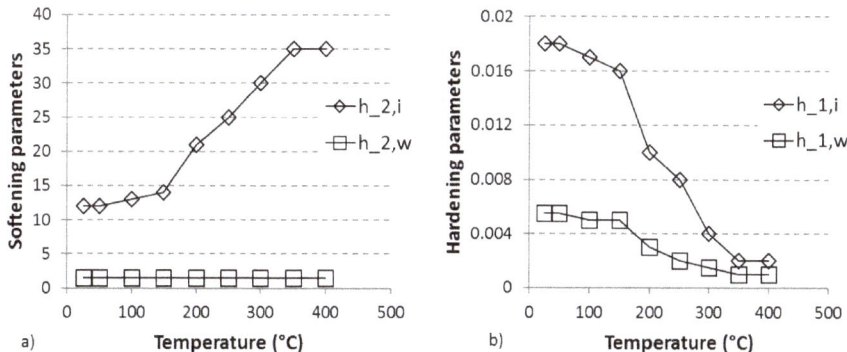

Figure 12. Softening (**a**) and hardening (**b**) parameters as a function of the temperature for dislocation densities walls (w) and interiors (i).

The evolution of grain size after recrystallisation in samples deformed by plane strain compression can be explained as follows. As the critical bulge size for Strain Induced Boundary Migration has an inverse relationship to the stored energy of deformation, more grains can nucleate in the middle of the deformation crosses, where the accumulated strain is larger, than in regions of lower strain. The new grains impinge on each other upon growth and limit the grain size in the deformation crosses. Furthermore, the recrystallised grains are elongated in the direction of the alignment of the large aluminides. This is especially obvious during annealing at 300 °C between the deformation crosses. The grain growth in the transverse direction is severely limited in comparison to the rolling direction. In those regions where the intermetallic phases are located, the pinning effect on the moving boundaries increases accordingly, enough to stop them. At higher deformations and annealing temperatures, the same trend would be observed if the grains did not impinge on each other before reaching the bigger aluminides. The small aluminides, homogeneously distributed, might still slow down the boundary migration but cannot completely pin the grain boundaries during recrystallisation. Within the time frame of the measurements, the grain size appears to be stabilised. As grain growth after recrystallisation is driven only by capillarity effects, the pressure on the boundaries is much lower than during recrystallisation. Pinning by the homogeneously distributed smaller aluminides can then effectively counter balance this pressure and prevent any further boundary movement. Although above 350 °C, the Mg_2Si phase dissolves, it does not affect the recrystallisation kinetics, because these particles are too coarse and too widely spaced to effectively pin the boundaries. The effects of the finely dispersed aluminides and of the lines of coarse aluminides are dominant. The most critical parameter in determining the final grain size appeared to be density of nuclei. The more grains nucleate, the less space they must grow and the lower the final grain size. By slightly increasing the density of nuclei being created, the predicted grain radius can be decreased to match the experimental measurements.

Author Contributions: C.P., R.B., and P.L. conceived and designed the experiments; R.B., P.L., and S.M. performed the experiments; C.P., R.B., P.L., and M.S. analysed the data; R.B. conceived the model, and C.P. and P.S. modified it; C.P. and R.B. wrote the paper.

Funding: Financial support by the Austrian Federal Government (in particular from Bundesministerium für Verkehr, Innovation und Technologie and Bundesministerium für Wissenschaft, Forschung und Wirtschaft) represented by Österreichische Forschungsförderungsgesellschaft mbH and the Styrian and the Tyrolean Provincial Government, represented by Steirische Wirtschaftsförderungsgesellschaft mbH and Standortagentur Tirol, within the framework of the COMET Funding Programme.

Conflicts of Interest: The authors declare no conflict of interest.

References

1. Johnson, G.R.; Cook, W.H. A constitutive model and data for metals subjected to large strains, high strain rates and high temperatures. In Proceedings of the 7th International Symposium on Ballistics, Den Haag, The Netherlands, 19–21 April 1983.
2. Lin, Y.C.; Chen, Xi.; Liu, G. A modified Johnson–Cook model for tensile behaviors of typical high-strength alloy steel. *Mater. Sci. Eng. A* **2010**, *527*, 6980–6986. [CrossRef]
3. Fields, D.S.; Backofen, W.A. Determination of strain hardening characteristics by torsion testing. *Proc. Am. Soc. Test. Mater.* **1957**, *57*, 1259–1272.
4. Poletti, C.; Halici, D.; Xue, S. Flow Instabilities During Hot Deformation of Titanium Alloys and Titanium Matrix Composites. In Proceedings of the 13th World Conference on Titanium, San Diego, CA, USA, 16–20 August 2015. [CrossRef]
5. Kocks, G.I. Laws for Work-Hardening and Low-Temperature Creep. *J. Eng. Mater. Technol.* **1976**, *98*, 76–85. [CrossRef]
6. Estrin, Y. Dislocation-Density-Related Constitutive Modeling. In *Unified Constitutive Laws of Plastic Deformation*; Academic Press, Inc.: San Diego, CA, USA, 1996.
7. Roters, F.; Raabe, D.; Gottstein, G. Work hardening in heterogeneous alloys—A microstructural approach based on three internal state variables. *Acta Mater.* **2000**, *48*, 4181–4189. [CrossRef]
8. Nes, E. Modelling of work hardening and stress saturation in FCC metals. *Prog. Mater. Sci.* **1998**, *41*, 129–193. [CrossRef]
9. Humphreys, F.J.; Hatherly, M. *Recrystallization and Related Annealing Phenomena*, 2nd ed.; E. Ltd.: Oxford, UK, 2004.
10. Bellier, S.P.; Doherty, R.D. The structure of deformed aluminium and its recrystallization—Investigations with transmission Kossel diffraction. *Acta Metall.* **1977**, *25*, 521–538. [CrossRef]
11. Orowan, E. Problems of plastic gliding. *Proc. Phys. Soc.* **1940**, *52*, 8–22. [CrossRef]

12. Taendl, J.; Nambu, S.; Orthacker, A.; Kothleitner, G.; Inoue, J.; Koseki, T.; Poletti, C. In-situ observation of recrystallization in an AlMgScZr alloy using confocal laser scanning microscopy. *Mater. Charact.* **2015**, *108*, 137–144. [CrossRef]
13. Taylor, G.I.; Ma, F.S. Plastic strain in metals. 28th May lecture to the Institute of Metals. *J. Inst. Met.* **1938**, *67*, 307–324.
14. Taylor, G.I. The mechanisms of Plastic Deformation of Crystals. Part I.—Theoretical. *Proc. R. Soc. A* **1934**, *145*, 362–387. [CrossRef]
15. Hull, D.; Bacon, D.J. *Introduction to Dislocations*, 4th ed.; Butterworth-Heinemann: Oxford, UK, 2001.
16. Mecking, H.; Kocks, U.F. Kinetics of flow and strain-hardening. *Acta Metall.* **1981**, *29*, 1865–1875. [CrossRef]
17. Caillard, D.; Martin, J.L. *Thermally Activated Mechanisms in Crystal Plasticity*, 1st ed.; E. Ltd.: Oxford, UK, 2003.
18. Read, W.T.; Schockley, W. Dislocation Models of Crystal Grain Boundaries. *Phys. Rev.* **1950**, *78*, 275–289. [CrossRef]
19. Smith, C.S. Grains, Phases, and Interfaces—An Interpretation of Microstructure. *Trans. Metall. Soc. AIME* **1948**, *175*, 15–51.
20. Avrami, M. Kinetics of Phase Change: I General Teory. *Chem. Phys.* **1939**, *7*, 1103–1112. [CrossRef]
21. Tiryankioglu, M.; Staley, J.T. Physical Metallurgy and the Effect of Alloying Additions in Aluminium Alloys. In *Handbook of Aluminum; Physical Metallurgy and Processes*; Totten, G.E., MacKenzie, D.S., Eds.; CRC Press: Boca Raton, FL, USA, 2003; Volume 1, pp. 82–83.
22. Clausen, B. *Characterisation of Polycrystal Deformation by Numerical Modelling and Neutron Diffraction Measurements*; Technical University of Denmark: Lyngby, Denmark, 1997.
23. Kabliman, E.; Sherstnev, P. Integrated Modeling of Strength Evolution in Al-Mg-Si Alloys during Hot Deformation. *Mater. Sci. Forum* **2013**, *765*, 429–433. [CrossRef]
24. Zurob, H.S.; Bréchet, Y.; Dunlop, J. Quantitative criterion for recrystallization nucleation in single-phase alloys: Prediction of critical strains and incubation times. *Acta Mater.* **2006**, *54*, 3983–3990. [CrossRef]
25. Sevillano, J.G.; van Houtte, P.; Aernoudt, E. Large strain work hardening and textures. *Prog. Mater. Sci.* **1980**, *25*, 69–134. [CrossRef]
26. Poletti, C.; Rodriguez-Hortalá, M.; Hauser, M.; Sommitsch, C. Microstructure development in hot deformed AA6082. *Mater. Sci. Eng. A* **2011**, *528*, 2423–2430. [CrossRef]

© 2018 by the authors. Licensee MDPI, Basel, Switzerland. This article is an open access article distributed under the terms and conditions of the Creative Commons Attribution (CC BY) license (http://creativecommons.org/licenses/by/4.0/).

Article

Combined Calorimetry, Thermo-Mechanical Analysis and Tensile Test on Welded EN AW-6082 Joints

Philipp Wiechmann [1], Hannes Panwitt [2], Horst Heyer [2], Michael Reich [1,*], Manuela Sander [2] and Olaf Kessler [1,3]

1. Institute of Materials Science, Faculty of Mechanical Engineering and Marine Technology, University of Rostock, Albert Einstein-Str. 2, 18059 Rostock, Germany; philipp.wiechmann@uni-rostock.de (P.W.); olaf.kessler@uni-rostock.de (O.K.)
2. Institute of Structural Mechanics, Faculty of Mechanical Engineering and Marine Technology, University of Rostock, Albert Einstein-Str. 2, 18059 Rostock, Germany; hannes.panwitt2@uni-rostock.de (H.P.); horst.heyer@uni-rostock.de (H.H.); manuela.sander@uni-rostock.de (M.S.)
3. Competence Centre CALOR, Department Life, Light & Matter, Faculty of Interdisciplinary Research, University of Rostock, Albert-Einstein-Str. 25, 18059 Rostock, Germany
* Correspondence: michael.reich@uni-rostock.de; Tel.: +49-381-498-9490

Received: 11 July 2018; Accepted: 7 August 2018; Published: 9 August 2018

Abstract: Wide softening zones are typical for welded joints of age hardened aluminium alloys. In this study, the microstructure evolution and distribution of mechanical properties resulting from welding processes of the aluminium alloy EN AW-6082 (AlSi1MgMn) was analysed by both in-situ and ex-situ investigations. The in-situ thermal analyses included differential scanning calorimetry (DSC), which was used to characterise the dissolution and precipitation behaviour in the heat affected zone (HAZ) of welded joints. Thermo-mechanical analysis (TMA) by means of compression tests was used to determine the mechanical properties of various states of the microstructure after the welding heat input. The necessary temperature–time courses in the HAZ for these methods were measured using thermocouples during welding. Additionally, ex-situ tensile tests were done both on specimens from the fusion zone and on welded joints, and their in-depth analysis with digital image correlation (DIC) accompanied by finite element simulations serve for the description of flow curves in different areas of the weld. The combination of these methods and the discussion of their results make an essential contribution to understand the influence of welding heat on the material properties, particularly on the softening behaviour. Furthermore, the distributed strength characteristic of the welded connections is required for an applicable estimation of the load-bearing capacity of welded aluminium structures by numerical methods.

Keywords: AlMgSi alloy; EN AW-6082; welding; mechanical properties; microstructure; DSC; thermo-mechanical analysis; digital image correlation; tensile test; numerical simulation

1. Introduction

Wrought EN AW-6082 (AlSi1MgMn) alloy, as an age hardening aluminium alloy, has excellent weldability, corrosion resistance and mechanical strength and is widely used in the automobile and shipbuilding industries. The major alloying elements of this aluminium alloy 6082 are Mg and Si, which can increase the strength of the alloy through precipitation hardening. The welding of aluminium alloys can lead to defects such as porosity, incomplete fusion and hot cracking, and thus the welding work can be challenging. Age hardening aluminium alloys such as 6082, whose strength is increased by precipitation hardening, always exhibit phase transformation and a softening phenomenon because of the heat input generated during the welding process [1,2]. A proven method for investigations of such softening is the characterisation of microstructure and mechanical properties of welded joints by metallography as well as by standard load tests. Results (e.g., [3]) show the decreases of base material strength within the heat affected zone due to the dissolution of strengthening precipitates. For deeper knowledge and understanding of the softening phenomena, an in-situ characterisation of the microstructure development would be preferable. Differential scanning calorimetry is a suitable technique to record the precipitation and dissolution behaviour in situ during the heat treatment of aluminium alloys [4]. The method was initially developed for analysis of the precipitation behaviour during cooling after solution annealing and was subsequently expanded to the analysis of the short-term heat treatment of age-hardening aluminium alloys [5,6]. For a correct understanding of softening phenomena within the HAZ, knowledge of phase transformations during heating would be necessary. In this work, DSC was used for the first time to investigate the dissolution and precipitation behaviour of an age-hardened AlMgSi alloy when heated under typical temperature–time curves of a welding process. The results of the thermal analysis are discussed alongside the distributed mechanical properties of the HAZ, which have been determined in two ways. First, welded joints were investigated with elaborate load tests supported by numerical analysis. Second, the mechanical properties of a wide variety of microstructures caused by welding heat input were determined through thermo-mechanical analysis. The results of this work contribute to a better understanding of the development of mechanical properties in HAZ and make it possible to provide realistic material models for structure–mechanical investigations using the finite element method. In particular, the aim of the present project was to use the obtained results for the representation of the material characteristics of welded aluminium cross joints and to predict their limit load behaviour with numerical simulations.

2. Materials and Methods

2.1. Investigated Aluminium Alloy

The experiments of this study were performed on a wrought aluminium alloy, EN AW-6082 (BIKAR-Aluminium GmbH, Korbußen, Germany), which was supplied as a 10 mm thick plate in the initial state T651. According to DIN EN 515 the treatment T651 includes solution annealing, quenching, stretching by 1.5% to 3% and subsequent artificial aging. EN AW-4047 (MTC GmbH, Meerbusch, Germany) was used as welding filler material for welding specimens. The chemical composition of EN AW-6082 and fusion zone material of a butt joint determined with optical emissions spectroscopy (OES) is given in Table 1 in addition to the specifications from DIN EN 573-3 [7].

Table 1. Mass fraction of alloying elements in the investigated EN AW-6082 alloy, fusion zone material of a butt joint and weld filler material EN AW-4047, in percent.

Material/alloy	Source	Si	Fe	Cu	Mn	Mg	Cr	Zn
EN AW-6082	OES	0.83	0.38	0.06	0.48	0.92	0.03	0.01
EN AW-6082	DIN EN 573-3	0.7–1.3	$\leq$0.5	$\leq$0.1	0.4–1.0	0.6–1.2	$\leq$0.25	$\leq$0.2
EN AW-4047A	DIN EN 573-3	11–13	0.6	0.3	0.15	0.1	-	$\leq$0.2
Fusion zone material	OES	7.23	0.29	0.03	0.19	0.39	0.02	<0.01

Table 2 shows the mechanical properties in three different directions (rolling direction 0°, 45° and 90°, as in [8]) of the base material determined from tensile tests. A comparison with the standard shows that the properties of the base material fit or exceed the required values in all directions. The differences between the directions in the present rolled plate material are negligible compared to the differences in extruded material (e.g., Chen et al. [9]). Thus, isotropic behaviour can be assumed [10]. In this study, the mechanical properties from the 0°-specimens are used for the base material.

Table 2. Mechanical Properties of EN AW-6082 T651 depending on the rolling direction.

Rolling Direction	E (N/mm^2)	R_m (N/mm^2)	$R_{p0.2}$ (N/mm^2)	A_5 (%)
0°	70800	308	289	12.2
45°	70000	303	278	13.0
90°	71300	308	284	11.5
Max. Difference	1.8%	1.5%	3.8%	11.8%
DIN EN 485-2 [11]	70000	300	255	9

Furthermore, for different methods investigating the material behaviour several different samples were used. Table 3 gives an overview of the different specimen geometries and dimensions.

Table 3. Overview on used samples.

Method	Previous Treatment	Geometry	Dimensions in mm
Temperature measurement	Initial state	T-joint *	240 × 160 × 10 plus 240 × 71 × 10
DSC, heat flow	Initial state	Cylindrical	Ø6 × 21.65
DSC, power compensated	Initial state	Cylindrical	Ø6.4 × 1
TMA	Initial state	Cylindrical	Ø5 × 10
Tensile tests	Initial state	Cylindrical **	Ø8 × 48
Tensile tests	Butt welded	Cylindrical **	Ø6 × 36
Tensile tests, DIC	Butt welded	Flat specimen **	25 × 6 (B×T), smooth, R40, R10

* see Figure 2, ** see Figure 4.

The high strengths in aluminium alloys are achieved in particular by precipitation hardening [12]. The precipitation sequence of Al-Mg-Si alloys was described by Edward and Dutta et al. [13,14]. An overview of these precipitates with information on dimensions, coherence, shape and further remarks was given by Polmear [15]. In Al-Mg-Si alloys, the beta phase results in maximum strengths [13].

The precipitation behaviour of several Al-Mg-Si alloys during cooling was investigated with DSC and microstructure analysis (optical microscopy (OM), SEM and TEM) [16–19]. Two different reaction areas, high (HTR) and low temperature reactions (NTR), were detected. In part, there is also a third middle temperature reaction (MTR). The high temperature reactions were correlated with the precipitation of Mg$_2$Si and the low temperature reactions of the precipitation of precursor phases. Precipitation behaviour depend strongly on initial state and chemical composition The critical cooling rate of 6082 can vary by factor of 10 depending on Mg and Si content [20].

In [21], the precipitation behaviour of the same batch of 6082 in the same initial state as in this study was analysed depending on different annealing conditions. The precipitation behaviour depends above all on whether there is a complete or incomplete dissolution of secondary particles at the onset of cooling.

The dissolutions and precipitations of Al-Mg-Si alloys during heating were also analysed with DSC and it was linked to the mechanical properties by TMA [5,21,22]. Osten et al. [21] investigated the dissolution and precipitation behaviour of several Al-Mg-Si alloys, including 6082, in various initial states during heating and has assigned the measured peaks to specific reactions through extensive literature research.

2.2. Welding Procedure and Temperature Measurements

Considering the aim of the project, butt welded joints and T-joints were used (see Figures 1 and 2), which were processed manually with metal inert gas welding (MIG) with three and four beads, respectively. Plates of EN AW-6082 T651 were welded with EN AW-4047 (wire diameter 1.2 mm) as weld filler material. Welding was conducted with direct current and positive polarity. A mixture of argon and helium (70%/30%) was applied as shielding gas. A ceramic weld pool backing was used for all joints. Further welding parameters are listed in Table 4.

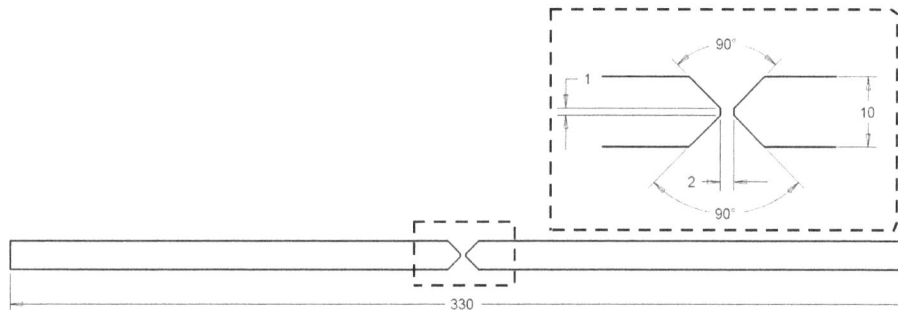

Figure 1. Prepared plates for butt welding, length of plates was 500 mm, lengths in mm.

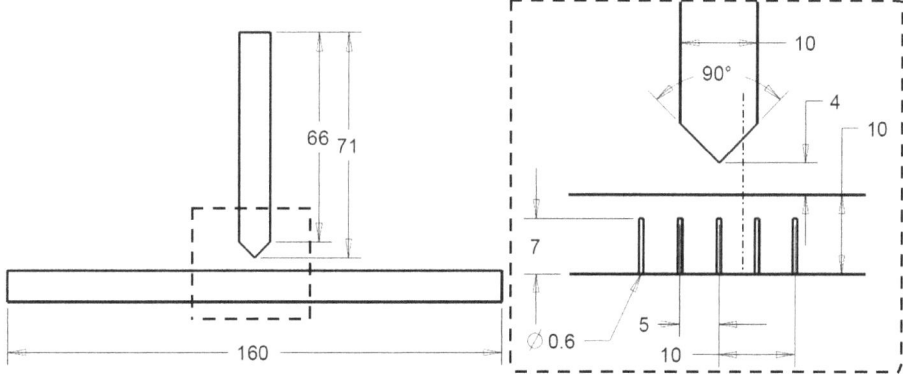

Figure 2. Sketch of prepared T-joint including drill holes for thermocouples, lengths in mm.

Table 4. Welding parameters of EN AW-6082 plates.

Joint	Welding Bead	Current (A)	Voltage (V)	Wire Feed (m/min)	Wire Diameter (mm)
Butt joint	1	145	23.5	7.5	1.2
	2 & 3	145	23.5	7.5	1.2
T-joint	1 & 2	204	24.4	9.5	1.2
	3 & 4	188	23.7	8.5	1.2

A temperature–time course in the heat-affected zone during a real welding process is needed as input data for differential scanning calorimetry and for thermo-mechanical analysis. Eight thermocouples (Type K, 0.5 mm, Therma Thermofühler GmbH, Lindlar, Germany) that were completely inserted in drilled holes simultaneously measured the temperature with a frequency of 50 Hz. The geometry of the prepared aluminium sheets is displayed in Figure 2 including the positions of the holes for thermocouples. The diameter of the holes was 0.6 mm, slightly larger than the diameter of thermocouple wire, to ensure that the thermocouples could be positioned at the end of drilled holes. The length of this T-joint was 240 mm and the thermocouple holes were drilled lengthwise at 80 and 160 mm from the edge.

2.3. Differential Scanning Calorimetry

The heating rate range of 0.01–5 K s^{-1} was investigated by direct DSC with two types of calorimeters: CALVET-type heat-flux DSC (DSC 121 and Sensys, Setaram, Caluire-et-Cuire, France) for slower (0.01–0.1 K s^{-1}) and power-compensated DSC for faster (0.3–5 K s^{-1}) scanning rates (Pyris Diamond and Pyris DSC 8500, PerkinElmer, Waltham, MA, USA). The samples used for heat-flux DSC had a cylindrical geometry with 6 mm diameter, 21.65 mm height and a mass of 1600 mg. Cylindrical samples with 6.4 mm diameter, 1 mm height and a mass of 80 mg were investigated in the power-compensated DSC devices. All experiments were carried out with an alloyed sample in one micro furnace and a pure–aluminium reference (99.9995% purity) with the same geometry in the other micro furnace. The samples and references were packed in pure-aluminium crucibles.

For investigation of very fast heating rates, which are typical for the HAZ during welding, direct DSC cannot be used, because the heating rate limit of the devices is exceeded. Instead, the indirect DSC method was used. Zohrabyan et al. [23] developed the differential reheating method to extend the temperature rate range. The schematic procedure of this method is shown in Chapter 3.2 together with its results. Rapid heating took place in the quenching dilatometer Bähr 805 A/D (BÄHR Thermoanalyse GmbH, Hüllhorst, Germany). The device is explained in Chapter 2.4. For indirect DSC, the samples had the same geometry (diameter of 6.4 mm, height of 1 mm, mass of 80 mg) as for direct DSC in the power-compensated devices. The samples were heated with rates from 20 to 100 K s^{-1} to temperatures of 200 °C to 450 °C, respectively, with an interval of 25 K. To preserve the state of the material at the maximum temperature, the samples were immediately quenched with maximum gas flow from He. After heat treatment, the samples were directly frozen at −80 °C until being reheated in the DSC device.

Reheating in the DSC device was performed with a scanning rate of 1 K s^{-1} to a maximum temperature of 575 °C.

The data processing of raw measured heat flow curves applied in this study was described in detail by Fröck et al. [24]. To obtain high-quality DSC results, the following sequence of experiments was conducted: sample measurement–baseline measurement, sample measurement. This is an efficient method to obtain a baseline for each sample measurement immediately. Baseline measurements were carried out with two pure aluminium references in the micro furnaces and the same temperature program as for the sample measurements. Baseline measurements were made to ascertain the current device specific curvature, which can change significantly within hours. This curvature is removed by subtracting the baseline determined in a timely manner.

The comparison of DSC curves of different sample masses m_s and scanning rates β requires a normalisation of the measured heat flow signal. For this reason, the specific heat capacity $c_{P_{excess}}$ [25] is calculated according to:

$$c_{P_{excess}} = \frac{\dot{Q}_s - \dot{Q}_{BL}}{m_s \beta} \ (\text{in J} \cdot \text{g}^{-1} \cdot \text{K}^{-1}) \quad (1)$$

with heat flow of baseline $\dot{Q}_{BL}$ and sample measurement $\dot{Q}_s$.

Remaining artefacts such as overshoots at the start and end of a scanning step were removed. The residual curvature of $c_{P_{excess}}$-curves can be compensated for with a polynomial fit. This was applied only for heating curves with scanning rates of 0.01 K s^{-1} and 0.03 K s^{-1}, because, for this data processing step, reaction free zones at low and high temperatures are necessary [4,21].

The slow heating experiments (0.01–0.1 K s^{-1}) consist of 4–6 sample measurements and 2–3 baselines. In the heating rate range of 0.3–5 K s^{-1}, eight sample measurements and four baselines were performed for each scanning rate. For indirect DSC, four sample measurements and two baselines for each maximum temperature and each heating rate were conducted. The average curves of these experiments are plotted in the diagrams. In total, more than 220 DSC experiments were performed.

2.4. Thermo-Mechanical Analysis and Hardness Testing

Thermo-mechanical analysis measures the deformation of a material under compression or tension as a function of temperature. To analyse the mechanical properties of the aluminium alloy 6082 T651 depending on the parameters of a thermal welding cycle, a thermo-mechanical analysis has been performed in the quenching and deformation dilatometer type Bähr 805 A/D. A schematic of the cylindrical compression sample inside the testing machine is shown in Figure 3. During the investigation, the specimens with geometrical dimensions of Ø 5 mm × 10 mm are heated inductively by the surrounding induction coil. An additional perforated inner coil was used for inert gas cooling. The temperature of the specimen was controlled with thermocouples spot-welded onto the specimen surface. The samples retrace the temperature–time profiles, which were measured in the HAZ during welding. The compression tests were carried out after seven days natural aging at about 20 °C with a deformation rate of 1 mm/s. Thereby, force–displacement curves were recorded. Every combination of heating rate and temperature was repeated three or four times and revealed a good reproducibility. The determined load-displacement diagrams were evaluated to flow curves representing true stresses and true strains.

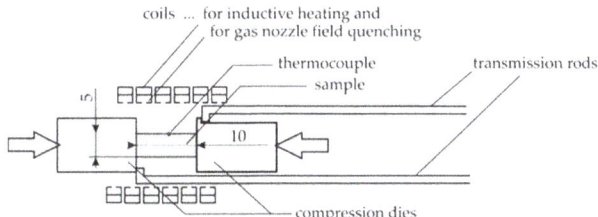

Figure 3. Schematic of cylindrical compression sample inside quenching and deformation dilatometer type Bähr 805 A/D.

During the evaluation of the compression tests, the absolute values of forces and displacements are calculated so that only positive strains and stresses are shown in the diagrams. These data can be compared directly with the results of the tensile tests.

For the hardness curve over the cross-section of weld seams, hardness values (HV1) were ascertained with the micro hardness tester HMV-2 from Shimadzu, Kyoto, Japan.

2.5. Tensile Tests on Welded Joints

To obtain the mechanical properties of the fusion zone (FZ), tensile tests on round specimens were conducted. The specimens had a diameter of 6 mm and were machined out of a V-shaped butt weld. Due to manufacturing limitations this specimen contained not only the weld material, but also small parts of the heat affected zone. Therefore, the results of these tests must be seen as integral values of the fusion zone and adjacent heat affected zone material. The displacements were measured by an extensometer.

To determine the mechanical behaviour of the heat affected zone, tensile tests on whole welded joints were conducted. Two plates of the base material were joined with an X-shaped butt weld, as described in Chapter 2.2. To obtain flat specimens the 10 mm thick welded plates were milled to 6 mm thickness (Figure 4a). Smooth and notched specimens (notch radius of 10 mm and 40 mm, Figure 4b) with a width of 25 mm in the smallest cross section were manufactured. Displacements and strains on the surface of the flat specimens were measured with a 2D digital image correlation system. Therefore, the surface of the specimens was prepared with a speckle pattern. The camera resolved the surface with a pixel size of 0.03 mm. The majority of the speckles had a size of 2 × 2 to 4 × 4 pixels. The data was processed with the software VIC 2D 6 (Correlated Solutions, Irmo, SC, USA).

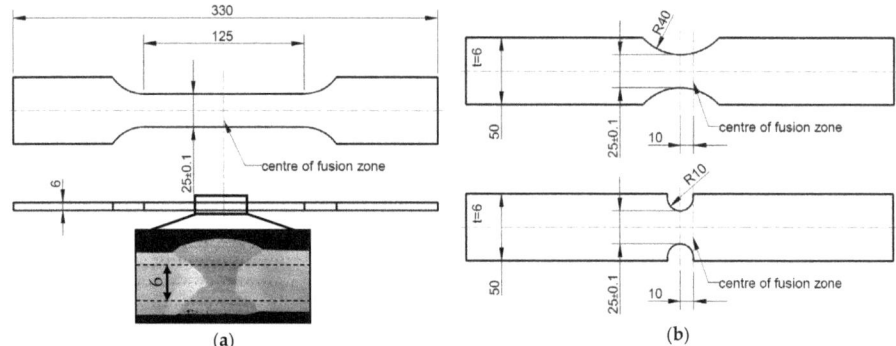

Figure 4. Geometries of the (**a**) smooth and (**b**) notched butt welded flat specimen for investigation of the HAZ, lengths in mm.

The DIC offers the possibility to place several virtual extensometers on the specimen surface with a freely chosen length. Therewith, force–displacement curves of several zones of the specimen can be obtained. The dimensions of the zones were first derived from hardness measurements and then compared with the DSC and TMA experiments.

For this investigation, the specimen was divided into four areas of interest: fusion zone and three areas in the heat affected zone (Z1, Z2 and Z3). Z3 was chosen such that differences with the base material (BM) are small and therefore the properties of the BM can be assumed (see hardness measurements in Chapter 3.3). Necking and fracture of the specimen was expected in Z1. Z2 filled the area between Z1 and Z3.

With the virtual extensometers, the force–displacement curves of each individual material zone were measured along with a global force–displacement curve including all zones. For the global curve, the virtual extensometer had a base length of 65 mm, whereas the extensometers of Z1 and Z2 were applied over the whole zone length of 12 mm and 9 mm, respectively. Accurate strain measurements can be achieved with the virtual extensometers due to their undeformed length of at least 300 pixels. Material properties from Z1 and Z2 can be obtained with this method without manufacturing separate specimens.

2.6. Determination of True Stress–Strain Curves

Tensile tests can be evaluated to determine the flow curve of a material. As long as uniform elongation occurs in the tests, the flow curve (equivalent von Mises strain σ_{vM} over total equivalent plastic strain ε_{pl}) can be calculated analytically as follows. First, the true strain ε

$$\varepsilon = \ln(1 + \varepsilon_e) \qquad (2)$$

and true stress σ

$$\varepsilon = \sigma_e(1 + \varepsilon_e) \qquad (3)$$

can be calculated from the engineering stresses σ_e and engineering strains ε_e. In this case, the true stress equals the von Mises equivalent stress σ_{vM}. The plastic strain ε_{pl} can be calculated by

$$\varepsilon_{pl} = \varepsilon - \frac{\sigma}{E} \qquad (4)$$

After onset of necking of the specimen, the stress state is not uniaxial anymore. To obtain the flow curves beyond the onset of necking, there are several analytical approaches. One often used method is to fit the values obtained by Equations (3) and (4) with a simple power law of the form

$$\sigma_{vM} = K\varepsilon_{pl}^n \qquad (5)$$

Another possibility is to calculate the parameters K and n of Equation (5) with the true stresses σ_m and plastic strains ε_m at the beginning of necking. The power law becomes

$$\sigma_{vM} = \sigma_m \left(\frac{\varepsilon_{pl}}{\varepsilon_m}\right)^{\varepsilon_m} \text{ for } \varepsilon_{pl} \geq \varepsilon_m \qquad (6)$$

and allows an extrapolation of the experimental data beyond the onset of necking.

However, neither method considers experimental results after the start of necking. Therefore, numerical simulations were conducted with the finite element program MarcMentat2013 to obtain flow curves with an iterative procedure. On the one hand, round specimens were simulated with rotational symmetric half models. On the other hand, 3D volume models were used to simulate flat specimens. In contrast to the geometry of the specimens, the deformation of the welded specimens is not symmetric in the tension direction due to strain localisation in the HAZ at one side of the fusion zone. Therefore, a quarter model with symmetry in width and thickness directions was used.

In this iterative procedure, the flow curve of the material is changed in a way that the resultant force–displacement curve in the simulation equals the force–displacement curve of the experiment. The detailed procedure was described by Gannon [25].

This method can be used for the base material and fusion zone material, since specimens with homogenous behaviour are assumed. For the HAZ, this method is not useable without modification, because the flat specimens do not consist of a homogeneous material (see Figure 4a). Furthermore, no necking or failure occurs in the Z2. Accordingly, the experimental stress–strain curve of the Z2 does not reach the tensile strength for this zone and σ_m and ε_m are unknown. Thus, the experimental result for the flow curve of the Z2 is extrapolated with a fitted power law given in Equation (5).

The flow curve of the Z1 can be obtained by iteration, but instead of using one single material the whole specimen with FZ, Z1, Z2 and Z3 (assumed properties of the BM) and their respective flow curves was modelled. The simulation of a complete specimen ensures that the edges of the Z1 behave correctly, because the different strengths of the adjacent Z2 and fusion zone hinder the deformation in width direction.

3. Results and Discussion

3.1. Temperature–Time Course in Heat Affected Zone

The cross section of the welded joint, which was used for temperature measurements, is shown in Figure 5 including thermocouple bores. The thermocouple wires were located at the end of the blind holes, so the distance between each weld bead and the points of temperature measurement was determined with these cross-section images.

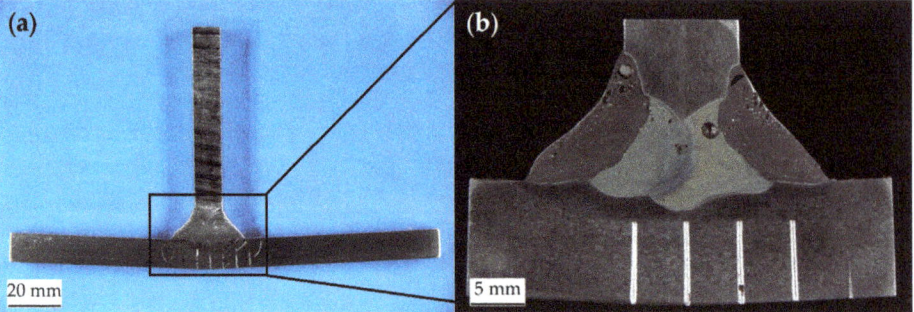

Figure 5. (a) Cross-section of welded T-joint; and (b) macro image of bores for thermocouples.

A typical temperature–time course in HAZ during welding and its three analysed parameters (heating rate, T_{max}, and cooling rate) are shown in Figure 6a. The heating in all recorded courses was nearly linear over a wide temperature range. The maximum temperature (T_{max}) was reached without a holding time and the cooling started immediately with a Newtonian course. Below 200 °C, the temperature decreases very slowly due to the relative small dimensions of joined plates, which heated up significantly. Therefore, only the cooling between T_{max} and 200 °C was used to calculate the mean cooling rate.

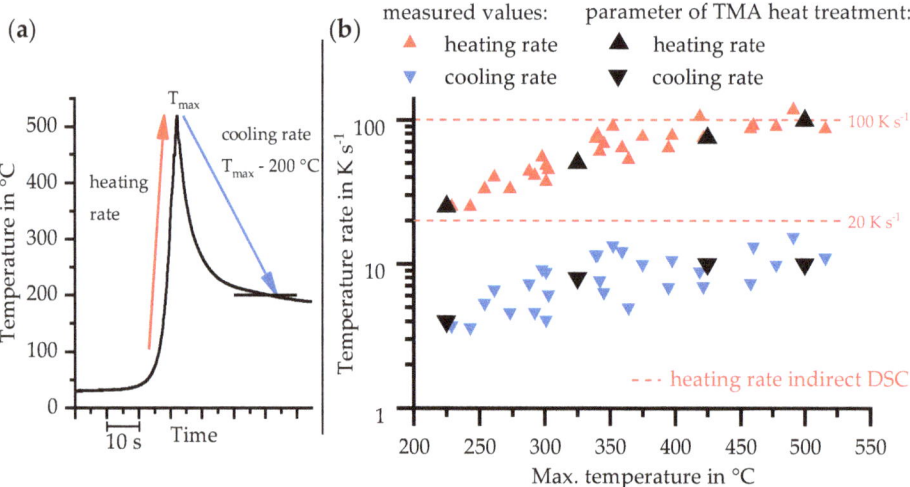

Figure 6. (a) Typical measured temperature–time course in HAZ; and (b) heating and cooling rates in the HAZ during MIG welding of EN AW-6082 depending on the maximum temperature and the resulting parameter of TMA heat treatment as well as the heating rates for indirect DSC.

The analysed heating and cooling rates in the HAZ during welding are plotted against T_{max} in Figure 6b. In principle, the heating and cooling rate increase as the maximum temperature rises, although a scattering of measured values occurs.

The three analysed parameters of temperature measurement revealed:

- Linear heating rates: 25–118 K s^{-1}
- Maximum temperatures (T$_{max}$): 229–516 °C
- Averaged cooling rates between T$_{max}$ and 200 °C: 3.5–15 K s^{-1}.

Because the maximum temperature correlates with distance from the fusion zone, these results are also plotted against the distance to weld bead in Figure 7.

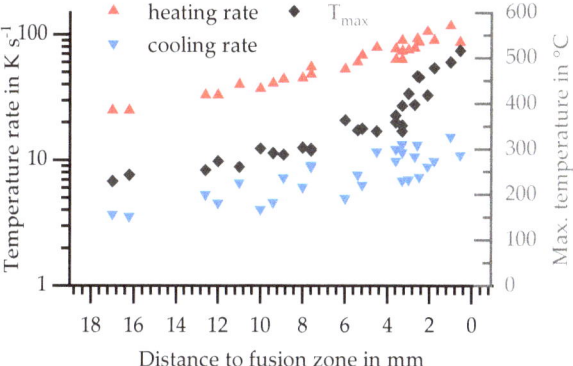

Figure 7. Parameters of temperature–time course dependent on distance to the weld seam.

These results, temperature rates and corresponding maximum temperatures, retrace different positions in the HAZ and were selected as parameters for TMA in this study. They are marked with black symbols in Figure 6b and given in Table 5. The chosen heating rates of indirect DSC (20 K s^{-1} and 100 K s^{-1}) are in the minimum and maximum range of these values.

Table 5. TMA parameters retracing HAZ.

Distance to Fusion Zone	Max. Temperature in °C	Heating Rate in K s^{-1}	Cooling Rate in K s^{-1}
Ca. 2 mm	500	100	10
Ca. 4 mm	425	75	10
Ca. 8 mm	325	50	8
Ca. 16 mm	225	25	4

3.2. Precipitation and Dissolution Behaviour of EN AW-6082 T651 in a Wide Dynamic Range

The excess heat capacity curves of heating the alloy EN AW-6082 with initial state T651 over a heating rate range from 0.01 K s^{-1} to 5 K s^{-1} up to 585 °C are plotted in Figure 8. During heating of aluminium alloys, dissolution and precipitation reactions occur. Precipitations were measured as exothermic peaks and dissolution as endothermic peaks. These reactions are alternating and overlap each other. Thus, the DSC curves show only the resulting sum signal, and only the initial temperature of the first and the final temperature of the last reaction are true signals.

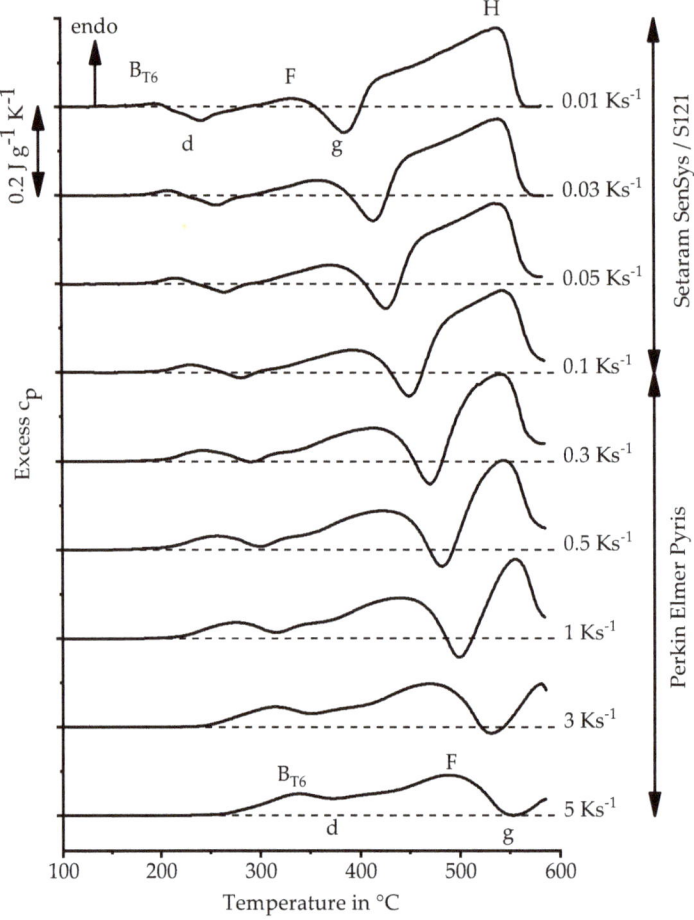

Figure 8. Direct DSC heating curves of EN AW-6082 T651 heating rates 0.01 K s^{-1} to 5 K s^{-1}.

The DSC curve recorded by Osten et al. [21] with another batch of EN AW-6082 with a 0.01 K s^{-1} heating rate resembles the curve from this study with the same scanning rate. There are only slight differences in reaction behaviour at slow scanning rates, which can be explained by differences in chemical composition, but the sequence of reactions is the same. Therefore, their interpretation of the reaction sequence is used in this study. The reactions were labelled here with the same characters [21].

The first peak B for the initial state T6 is induced by the dissolutions of GP-zones and β″, with β″ being the phase which effects the maximum strengths of Al-Mg-Si alloys [13]. The peak d corresponds to either the precipitation of β″ or β′ depending on initial state [13,15]. For the initial state T651, there is probably only a precipitation of β′, because β″ is already dissolved in the previous reaction. The reactions which cause the peaks F and g belong to the dissolution of β′ and the precipitation of β (Mg$_2$Si). The dissolution of the remaining precipitations, especially β (Mg$_2$Si), is recorded as final peak H. At very slow heating rates, there is a reaction-free range following peak H, which indicates a complete dissolution of these particles [21].

As the heating rates increase, there is a shift of reactions to higher temperatures, which also results in an incomplete dissolution with fast heating. Furthermore, the curves shift in the endothermic direction. However, it is unlikely that dissolution will increase at faster heating rates. Rather, it can be assumed that the shift is thus caused because precipitation reactions are significantly more suppressed than dissolution reactions.

Figure 9a displays the temperature–time course of the indirect DSC method. To maintain the condition at T_{max} and to prevent quench-induced precipitation, quenching is performed with maximum gas flow after the first heating. The average cooling rates β of the Newtonian cooling course, which depended on the temperature interval considered, are listed in Table 6.

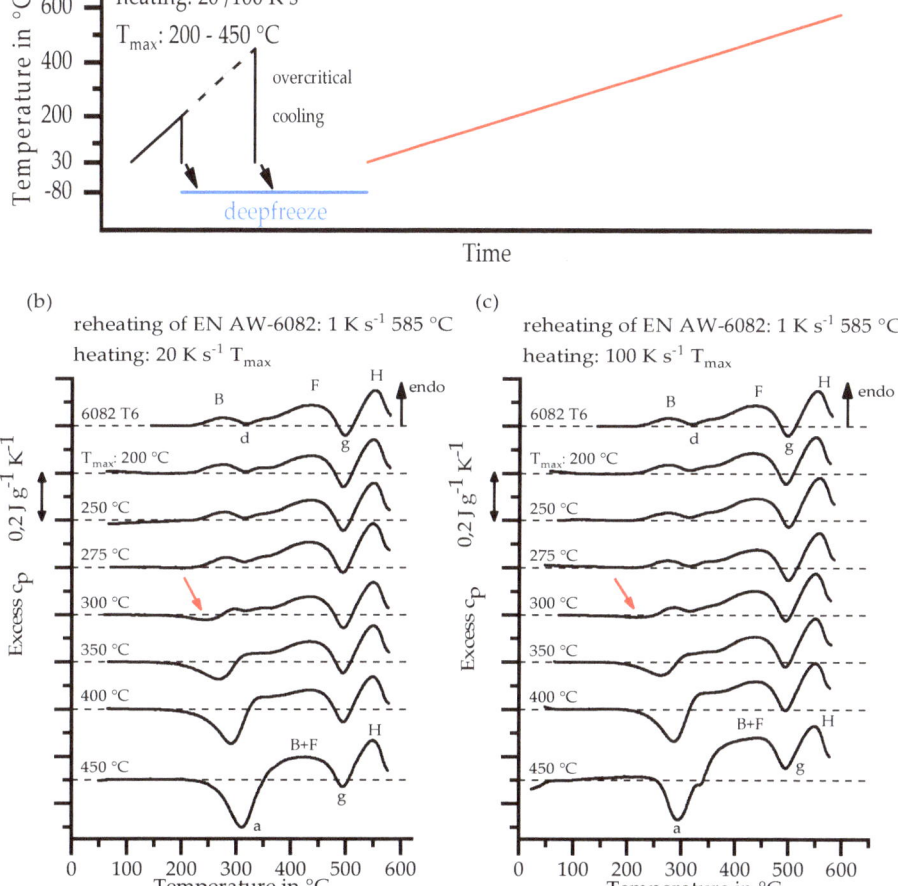

Figure 9. Indirect DSC: (**a**) schematic temperature–time course; and reheating DSC curves of heating rates: (**b**) 20 K s^{-1}; and (**c**) 100 K s^{-1}.

Table 6. Average cooling rates of heat treatment for indirect DSC.

Upper Temperature	Lower Temperature	Average Cooling Rate
450 °C	100 °C	~200 K s^{-1}
200 °C	100 °C	~120 K s^{-1}
100 °C	30 °C	>32 K s^{-1}

Fröck et al. [24] used the same batch of 6082 to investigate the influence of different solution conditions on the precipitation behaviour during subsequent cooling. For an incomplete solution state (after 540 °C for 1 min), the upper critical cooling rate (uCCR) of 100 K s^{-1} was ascertained. The cooling rates of the heat treatment for indirect DSC are higher than this uCCR in temperature ranges above 100 °C. It can thus be assumed that no significant precipitation reactions took place during cooling and the state of the material reached at maximum temperature remains.

The reheating curves are shown in Figure 9b,c. The reaction peaks are given the same characters as in Figure 8. Low curvature is present in the curves, which can give reasons for slight quantitative differences between single curves. This is particularly apparent at higher temperatures, e.g., the peaks g and H, or the slope of reaction free zone are influenced by this remaining curvature. Nevertheless, the development of reactions is clearly visible. The reheating curves of the investigated heating rates 20 K s^{-1} and 100 K s^{-1} show no significant differences for the same T_{max}. Depending on T_{max}, there is a substantial development in the reheating curves for each heating rate. In conclusion, the reactions taking place in the HAZ are mainly dependent on T_{max} and are less dependent on the heating rate, at least in the investigated range.

The reheating curves from the initial state EN AW-6082 T651 to T_{max} of 275 °C are almost identical. That means no significant reactions take place until heating to this temperature. From T_{max} 300 °C an exothermic reaction starts (see arrows in Figure 9b,c). These reaction peaks increase with a higher maximum temperature of first heating. During the first heating, existing precipitates are dissolved increasingly with rising temperature. A supersaturation occurs due to overcritical cooling, which causes the measured precipitation reactions during reheating. This dissolution reaction B_{T651} during rapid heating is crucial for softening in the HAZ.

The reaction peaks determined by direct DSC and the dissolution reaction B_{T651} determined by indirect DSC are plotted in temperature–time courses of investigated heating experiments, to create a continuous heating dissolution diagram for a wide range of heating rates, as shown in Figure 10.

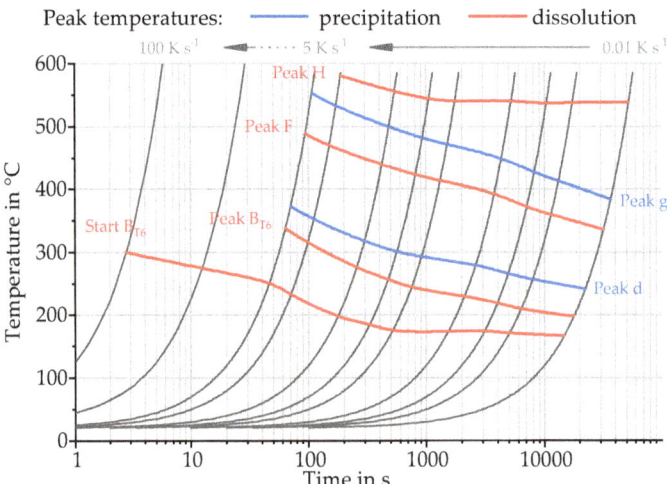

Figure 10. Continuous heating dissolution diagram EN AW-6082 T651 heating rates 0.01 K s^{-1} to 100 K s^{-1}.

The temperatures of dissolution or precipitation reactions during heating of EN AW-6082 T651 within a range of 0.01 K s^{-1} to 100 K s^{-1} can be taken from this diagram. For heating of 20 K s^{-1} to 100 K s^{-1}, investigated with indirect DSC, only the start of the dissolution reaction B_{T651} can be determined at temperatures between 275 °C and 300 °C.

3.3. Mechanical Properties of the HAZ

The results of hardness tests in Figure 11 provide an overview of properties as a function of distance to the weld centre. At a distance from the weld centre of more than 50 mm a constant hardness of about 100 HV1 was measured in the base material 6082 T651. At about 40 mm, a maximum hardness of 110 HV1 is reached. One reason for the increase in hardness may be that the initial state T651 was slightly underaged and the welding heat causes artificial ageing at this point. With decreasing distance, the hardness decreases significantly to a minimum of about 60 HV1. The hardness increases in the direct vicinity of the FZ. Hardness of the FZ was about 70–80 HV1.

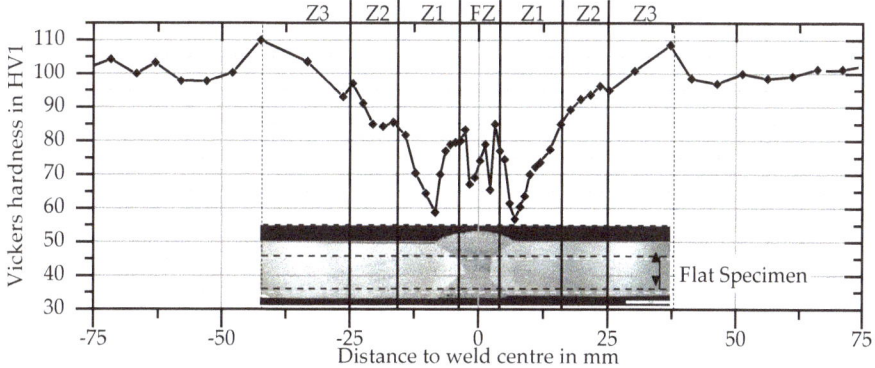

Figure 11. Hardness after welding and natural aging in plate centre.

In Figure 12, the results of TMA with parameters according to Table 5 are plotted against T_{max} for the short term heat treatment. The yield strength has been measured after seven days of natural ageing. Compared with the initial state, there is a small increase for T_{max} 225 °C. From T_{max} 225 °C to 425 °C, the yield strength decreases by about half to less than 130 N/mm². For the highest investigated T_{max} of 500 °C the yield strength increases slightly.

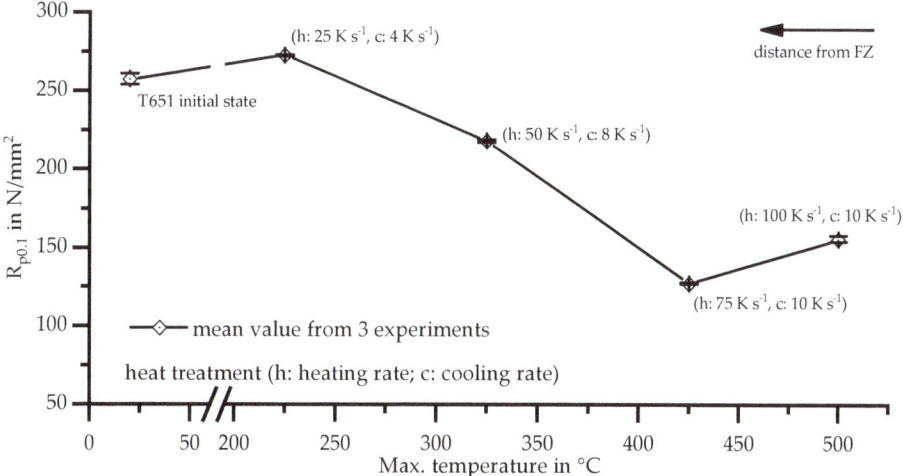

Figure 12. Yield strength after welding cycle and seven days natural aging depending on maximum temperature of short term heat treatment.

Microstructure analyses (SEM and TEM) were performed by Fröck et al. [24] with the same material after annealing at different maximum temperatures. During annealing, both complete and incomplete dissolution of secondary phase particles was achieved depending on the maximum temperature. As Figure 9 shows, there will be an incomplete dissolution for fast heating rates. In consideration of the quasibinary phase diagram Al-Mg$_2$Si [15], the same phases are expected after the TMA welding heat treatments as after solution annealing at 540 °C [24].

Because maximum temperature correlates with distance to the FZ, the course of the yield strength (Figure 12) depending on maximum temperature is similar to the hardness profile (Figure 11).

Regarding DSC and TMA, the HAZ of 6082 T6 can be divided in four areas.

A. Above 425 °C, solution annealing takes place. Rapid quenching near the FZ causes a supersaturated solid solution with potential for age hardening. Yield strength increases again after natural aging.
B. From 275 °C to 425 °C, β″ precipitates increasingly dissolve and yield strength decreases.
C. Weak precipitation of β″ happens at a temperature range of 225 °C, which leads to a slight increase in hardness and strength, but is hardly detected with DSC.
D. At a distance of more than 50 mm (below a certain T_{max}), the T6 state consisting of β″ precipitates remains nearly unchanged. Hardness is not affected.

3.4. Flow Curves in a Welded Joint

For the calculation of the flow curve of the base material and the fusion zone the engineering stress–strain curves determined from tensile tests on separate round specimens have been used. The mechanical properties of the fusion zone material were also determined from these tensile tests and are presented in Table 7. The chemical composition of the FZ according Table 1 appears in the range of cast aluminium alloys, which also roughly applies for its mechanical properties.

Table 7. Mechanical properties of the fusion zone material.

Material	E (N/mm^2)	R_m (N/mm^2)	$R_{p0.2}$ (N/mm^2)	A_5 (%)
FZ	71800	238	114	10

Whereas the base material shows ductile failure with necking after reaching the ultimate tensile strength, the fusion zone material fails without any noticeable necking (see Figure 13a). Therefore, the combined analytical and numerical approach described in Chapter 2.6 was used to calculate the flow curve of the base material. Numerical iterations were not necessary for the fusion zone material, since no necking and therefore no multiaxial stress state was present. The flow curve of the fusion zone was simply calculated by Equations (2)–(4). An extrapolation with Equation (6) extends the curve to a larger range of strains. To validate the obtained flow curves, a comparison between calculated and measured technical stress–strain curves is also shown in Figure 13a. No differences between the measured and simulated curves are visible.

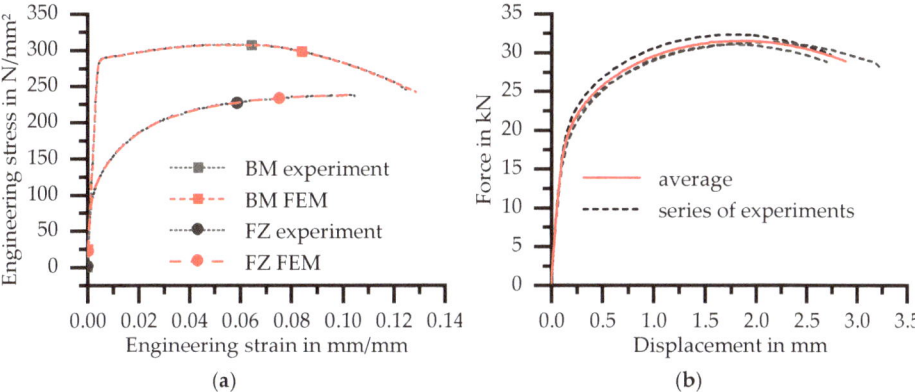

Figure 13. (a) Comparison between measured and calculated engineering stress–strain curves of base material and weld material; and (b) force–displacement curves of butt welded flat specimen.

Whereas all tests with the base and fusion zone material showed very good repeatability, the global force–displacement curves of the three tested welded flat specimens showed slight differences (see Figure 13b). It is assumed that the differences occur because of irregularities in the weld seam in length direction as well as due to specimen manufacturing from slightly different areas over the sheet thickness. To overcome the differences between curves, one average curve was used for comparison reasons with numerical simulations.

In addition to the global force–displacement curve, local force–displacement curves for the zones Z1 and Z2 were also determined by using the DIC. The respective lengths and positions of the material zones were derived from hardness measurements as shown in Figure 11. Z1 is the area between 4 mm and 16 mm distance to the centre of the fusion zone. This is the area in which fracture occurs during tensile tests. Z2 ends at 25 mm distance to the centre of the fusion zone when the hardness values increase to about 95% of the base material (i.e., about 95 HV1). For distances to the fusion zone larger than 25 mm (Z3), the properties of the unaffected base material are nearly reached.

For this arrangement, the experimental force–displacement data for Z2 only allows a calculation of the flow curve until about 0.3% plastic strain, because failure and strain localisation occurred in Z1. The curve of Z2 is extended to higher strains by fitting a power law according to Equation (5). The flow curve of Z1 is obtained afterwards through iteration with numerical simulations. In contrast to the base material, it was not possible to use Equations (2)–(4) until necking occurs (see Figure 14).

Due to the inhomogeneity of the HAZ, uniform elongation cannot be assumed until the maximum force is reached. Therefore, the experimental data were used as initial values for the numeric iteration only as long as agreement was maintained between the measured and calculated force–displacement curves.

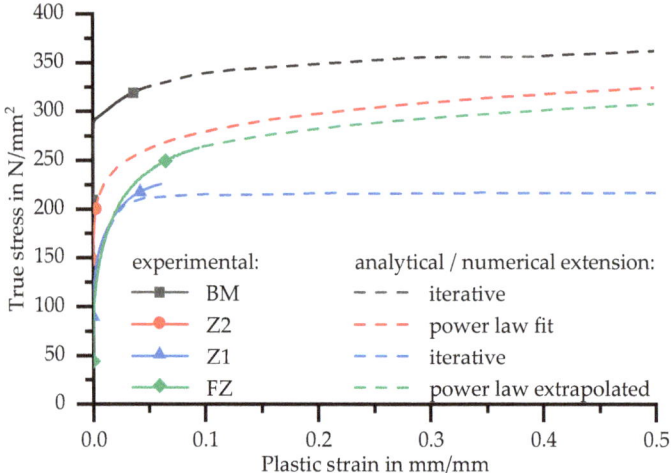

Figure 14. Flow curves of base and weld material, Z1 and Z2.

3.5. Validation of Obtained Flow Curves in the HAZ

The results of the tensile tests with butt welded flat specimen are here described in more detail. To validate the calculated curves, the strain distribution in the experiment (DIC) can be compared with the numerical results. Therefore, the maximum principal strain ε_1 was calculated in the DIC software at the specimen's surface. First, Figure 15 shows that no uniform elongation of the specimen is present even at low global displacements (maximum strain of 0.3%). Whereas the hardness measurements (see Figure 11) suggest the highest strain in Z1 next to the fusion zone, the fusion zone material dominates the deformation of the specimen at low strains. The behaviour of the flow curves (Figure 14) of the two zones explains this phenomenon: at low strains, the flow stress of the fusion zone material is less than the flow stress of Z1. A certain amount of strain hardening needs to occur for Z1 to dominate the deformation behaviour of the specimen.

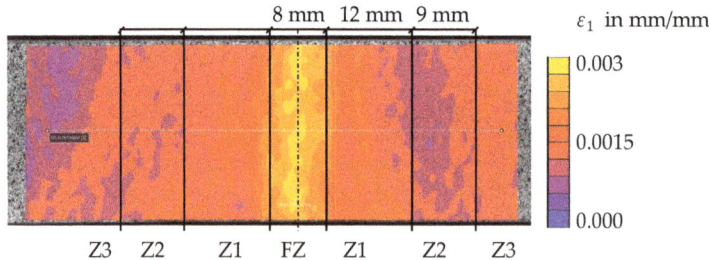

Figure 15. Strain distribution in a welded flat specimen at low global displacements (0.1 mm).

The top of Figure 16 shows the measured strain distribution of the specimen at 1.3 mm global displacement. In contrast to the strain distribution at low displacements, here, the highest strains occur almost symmetrically next to the fusion zone in Z1. For comparison, the bottom of Figure 16 shows the maximum principal strains calculated by the finite element (FE) simulation at the same displacement.

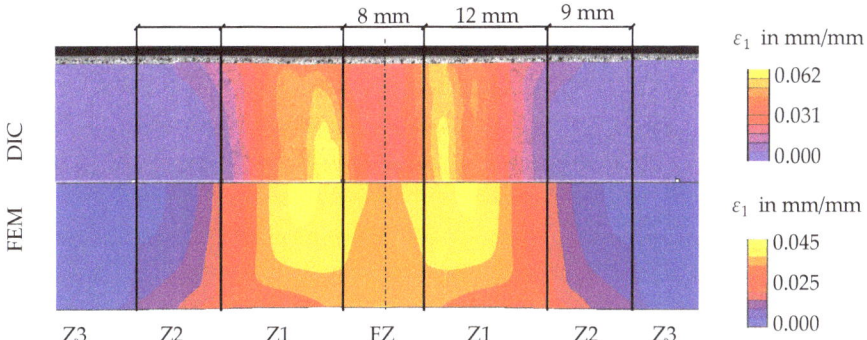

Figure 16. Comparison of the strain distribution in a butt welded flat tensile specimen at 1.3 mm global displacement: in the experiment (**top**); and in the FE simulation (**bottom**).

At first glance, the strain distribution shows good agreement between model and experiment. In both cases, the maximum strain is located in Z1. Whereas there are still noticeable strains in the fusion zone, the strain decreases within a few millimetres in Z2 to almost negligible strains in Z3. Since Z3 and Z2 deform less than Z1, the deformation of Z1 is constrained in the width direction. This constraint causes higher strains in Z2 at the edge of the specimen than in the middle. The constraining effect on the different material deformations becomes stronger in the simulation than in the experiment, because the FE model has no continuous change in material properties but rather an explicit change at the end of each material zone.

Another difference becomes visible by comparing the maximum strain values. The measured maximum strain is higher than in the numerical simulation and located closer to the fusion zone. It has to be pointed out that differences in maximum strain occur even though the measured and simulated force–displacement curves of the whole specimen are almost identical (see Figure 17). This is possible because the flow curve of Z1 averages a quite large area of the HAZ compared to high changes in hardness and the presumed mechanical properties in this zone. Since for example the lowest yield stress is averaged to a higher value, a smaller strain peak will be calculated.

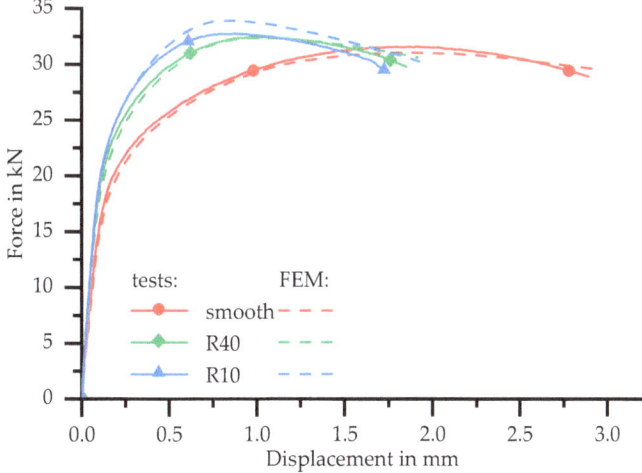

Figure 17. Force–displacement curves of notched and smooth butt welded flat specimen.

To investigate the behaviour of the HAZ in different multiaxial stress states and to validate the obtained flow curves in more detail, tensile tests and numerical simulations of notched specimens were conducted. Figure 17 shows a comparison of three different specimen shapes: smooth, large notch radius (40 mm) and small notch radius (10 mm) with equal nominal cross sections.

As it is well known, a notch will increase the maximum force: the smaller the notch radius, the higher the maximum force. The experimental results confirm this fact. However, the increase of the maximum force is small. This indicates that the inhomogeneity of the material dominates over the geometric effect due to the notch. The increase of maximum force is calculated by the FE simulations as well. However, the simulated and measured force–displacement curves of the specimen with large notch radii have good agreement, while the simulation overestimates the maximum force of the sharp notched specimen. Due to the material properties averaged in Z1, expressed by the flow curve, a larger force is required in the finite element simulation in order to map the local strain concentration in the notch root.

3.6. Correlation between Results of Tensile Tests and TMA of HAZ

When comparing the results of different methods, the type of joint used for temperature measurement (T-joint) and tensile test specimen (butt joint) must be considered. Whereas in a butt weld the heat can only be dissipated in two directions, the T-joints consists of three segments. Higher T_{max} as a function of distance and lower cooling rates can therefore be expected on the butt weld.

Hardness profiles (see Figure 18) of both welds were recorded in order to compare the welds and in particular the size of the HAZs with each other.

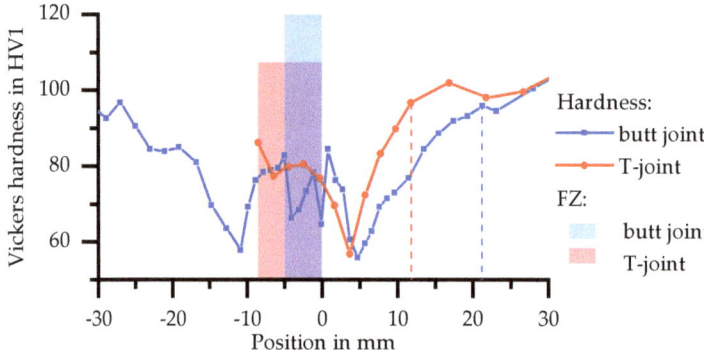

Figure 18. Hardness development in butt and T-joint depending on the position.

For the T-joint, the values of the FZ and the vertical plate are shown. In HAZ the hardness first decreases to a minimum, which is at a distance to FZ of 4 mm in the T-joint and at 5–6 mm in the butt joint. In further course the hardness increases until the initial value of about 100 HV1 is reached at 12 mm (T-joint) and 21 mm distance (butt joint) respectively. The courses of hardness are the same and the locations of, e.g., the minimum or initial hardness, match, considering the geometrically changed distribution of T_{max}.

Flow curves of the HAZ determined with tensile test and TMA can therefore be compared as seen in Figure 19.

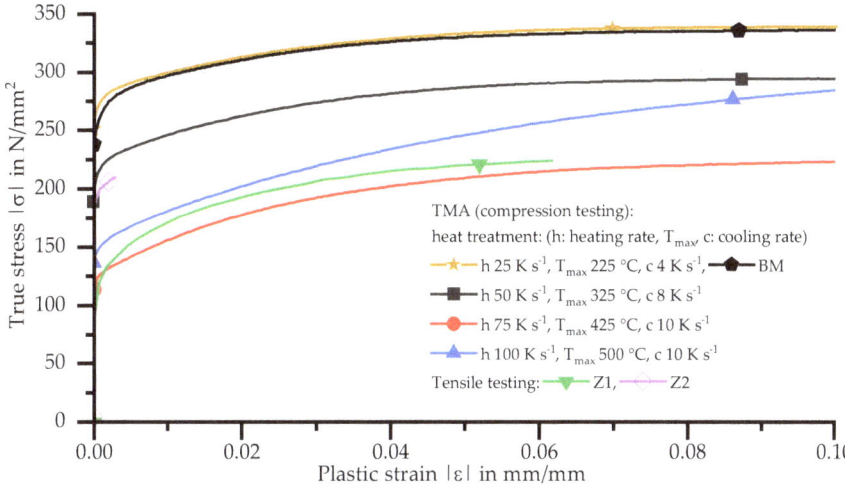

Figure 19. Flow curves of real and retraced HAZ determined with tensile and compression tests. The compression tests are carried out after the welding cycle and seven days of natural ageing.

The curve of Z1 is enveloped by the retraced HAZ with T_{max} 500 °C and 425 °C for low plastic strains. The experimentally determined flow curve of Z2 is only available up to the maximum stress of Z1, but it can be assumed that it follows the curve with T_{max} 325 °C. These results are in good agreement regardless of the different heat input in tensile and TMA samples. Whereas the tensile tests were carried out with specimens heated three times up to different maximum temperatures (three weld beads), the TMA specimens were subjected to a single, precisely defined short-term heat treatment.

The lowest mechanical properties are obtained by TMA with T_{max} 425 °C. By evaluating tensile tests in Z1 with the combined numerical and analytical approach described previously, the low mechanical properties as in the TMA cannot be determined due to the averaging in Z1. Averaging over the area of Z2 leads to a curve similar to the TMA flow curve with T_{max} 325 °C, even though no significant difference to the BM is expected at the end of Z2. As Figure 12 shows, the short-term heat treatment with T_{max} 225 °C results in maximum strength just above that of the BM.

It becomes visible that a combined approach with tensile tests, DIC and numerical simulations and DSC and TMA following temperature profiles during welding can lead to an improved description of material behaviour in different areas of a weld for a specific welding process and geometry. With knowledge of the maximum temperatures depending on the distance to the FZ, phase transformations obtained by DSC and material properties obtained in a TMA can therefore deepen understanding of the microstructural changes in the HAZ and refine numerical structure–mechanical simulations of welded components.

4. Conclusions

In this study, HAZ properties of welded joints made of AlMgSi wrought alloy EN AW-6082 T651 were investigated using several combined methods. The following conclusions can be drawn from the consideration of the individual results and their mutual discussion:

1. Dissolution and precipitation reactions in different areas of the HAZ can be analysed in situ with DSC. To retrace the thermal history in the HAZ, the temperature rate range was extended with the indirect DSC method.
2. There is a good agreement between results of phase transformations determined with DSC and changes in mechanical properties measured with TMA.

3. The softening in HAZ is strongly dependent on peak temperature. With increasing peak temperature, the initial state is increasingly dissolved and the material is softened to a minimum. Near the FZ, the mechanical properties increase due to strong dissolution of alloying elements and the associated potential for natural aging.
4. The development of dissolution and precipitation can be described by continuous heating dissolution diagrams, similar to welding-transformation diagrams of steels.
5. Mechanical properties from TMA and results of tensile tests on welded joints show good agreement in relevant HAZ zones.
6. Flow curves of the base material, fusion zone material and two areas in the HAZ in a butt welded joint can be calculated with a combined numerical and analytical approach. DIC measurements can provide the necessary force–displacement data in the HAZ without manufacturing separate specimens.
7. A certain amount of strain hardening in the FZ needs to occur before the HAZ dominates the deformation of welded flat specimen.
8. Small increases of maximum force with decreasing notch radii indicate that the inhomogeneity of the HAZ dominates over the geometric effect due to the notch.
9. Numerical simulations of notched tensile specimens with these flow curves lead to accurate force–displacement curves for large notch radii. The maximum principal strain is underestimated by numerical simulations, because of the averaging of material behaviour in the HAZ.

Author Contributions: All authors conceived and designed the experiments; P.W. and H.P. performed the experiments and analysed the data; and P.W., H.P. and Reich wrote the paper.

Funding: The authors acknowledge funding of this work from the German Research Foundation (DFG), within the scope of the research project "Materials Based Simulation of Limit Load Behaviour for Welded Aluminium Structures" (DFG RE3808/2-1 & DFG SA 960/7-1).

Conflicts of Interest: The authors declare no conflict of interest.

References

1. Collette, M.D. The impact of fusion welds on the ultimate strength of aluminum structures. In Proceedings of the 10th International Symposium on Practical Design of Ships and other Floating Structures, Bowie, MD, USA; 2007; pp. 944–952.
2. Missori, S.; Sili, A. Mechanical behavior of 6082-T6 aluminium alloy welds. *Metall. Sci. Technol.* **2000**, *18*, 12–18.
3. Mathers, G. *The Welding of Aluminium and Its Alloys*; CRC Press: Boca Raton, FL, USA, 2002; p. 34. ISBN 0-8493-1551-4.
4. Milkereit, B.; Kessler, O.; Schick, C. Precipitation and Dissolution Kinetics in Metallic Alloys with Focus on Aluminium Alloys by Calorimetry in a Wide Scanning Rate Range. In *Fast Scanning Calorimetry*; Schick, C., Mathot, V., Eds.; Springer International Publishing: Cham, Switzerland, 2016; Volume S, pp. 723–773. ISBN 9783319313290.
5. Fröck, H.; Graser, M.; Reich, M.; Lechner, M.; Merklein, M.; Kessler, O. Influence of short-term heat treatment on the microstructure and mechanical properties of EN AW-6060 T4 extrusion profiles: Part A. *Prod. Eng.* **2016**, *10*, 383–389. [CrossRef]
6. Graser, M.; Fröck, H.; Lechner, M.; Reich, M.; Kessler, O.; Merklein, M. Influence of short-term heat treatment on the microstructure and mechanical properties of EN AW-6060 T4 extrusion profiles—Part B. *Prod. Eng.* **2016**, *10*, 391–398. [CrossRef]
7. DIN Deutsches Institut für Normung e. V. 573-3. *Aluminium und Aluminiumlegierungen—Chemische Zusammensetzung und Form von Halbzeug*; DIN: Berlin, Germany, 2009.
8. Wang, T.; Hopperstad, O.S.; Larsen, P.K.; Lademo, O.-G. Evaluation of a finite element modelling approach for welded aluminium structures. *Comput. Struct.* **2006**, *84*, 2016–2032. [CrossRef]
9. Chen, Y.; Clausen, A.H.; Hopperstad, O.S.; Langseth, M. Stress–strain behaviour of aluminium alloys at a wide range of strain rates. *Int. J. Solids Struct.* **2009**, *46*, 3825–3835. [CrossRef]

10. Törnqvist, R. Design of Crashworthy Ship Structures. Ph.D. Thesis, Department of Mechanical Engineering, Technical University of Denmark, Lyngby, Denmark, 2003.
11. *Aluminium und Aluminiumlegierungen—Bänder, Bleche und Platten—Teil 2: Mechanische Eigenschafte, DIN EN. 485-2*; Deutsches Institut für Normung e. V.: Berlin, Germany, 2007.
12. Ostermann, F. *Anwendungstechnologie Aluminium*; Springer: Berlin/Heidelberg, Germany, 2014; ISBN 9783662438060.
13. Edwards, G.A.; Stiller, K.; Dunlop, G.L.; Couper, M.J. The precipitation sequence in Al-Mg-Si alloys. *Acta Mater.* **1998**, *46*, 3893–3904. [CrossRef]
14. Dutta, I.; Allen, S.M. A calorimetric study of precipitation in commercial aluminium alloy 6061. *J. Mater. Sci. Lett.* **1991**, *10*, 323–326. [CrossRef]
15. Polmear, I.J. *Light Alloys. From Traditional Alloys to Nanocrystals*, 4th ed.; Elsevier Butterworth-Heinemann: Amsterdam, The Netherlands, 2006; ISBN 0750663715.
16. Milkereit, B.; Kessler, O.; Schick, C. Recording of continuous cooling precipitation diagrams of aluminium alloys. *Thermochim. Acta* **2009**, *492*, 73–78. [CrossRef]
17. Milkereit, B.; Beck, M.; Reich, M.; Kessler, O.; Schick, C. Precipitation kinetics of an aluminium alloy during Newtonian cooling simulated in a differential scanning calorimeter. *Thermochim. Acta* **2011**, *522*, 86–95. [CrossRef]
18. Milkereit, B.; Wanderka, N.; Schick, C.; Kessler, O. Continuous cooling precipitation diagrams of Al-Mg-Si alloys. *Mater. Sci. Eng. A* **2012**, *550*, 87–96. [CrossRef]
19. Milkereit, B.; Starink, M.J. Quench sensitivity of Al-Mg-Si alloys: A model for linear cooling and strengthening. *Mater. Des.* **2015**, *76*, 117–129. [CrossRef]
20. Milkereit, B.; Schick, C.; Kessler, O. Continuous Cooling Precipitation Diagrams Depending on the Composition of Aluminum-Magnesium-Silicon Alloys. In Proceedings of the 12th International Conference on Aluminium Alloys, Yokohama, Japan, September 2010; Volume S, pp. 407–412.
21. Osten, J.; Milkereit, B.; Schick, C.; Kessler, O. Dissolution and precipitation behaviour during continuous heating of Al-Mg-Si alloys in a wide range of heating rates. *Materials* **2015**, *8*, 2830–2848. [CrossRef]
22. Milkereit, B.; Osten, J.; Schick, C.; Kessler, O. Continuous Heating Dissolution Diagrams of Aluminium Alloys. In *Proceedings of the 13th International Conference on Aluminum Alloys (ICAA13), 2012*; Weiland, H., Rollett, A.D., Cassada, W.A., Eds.; TMS (The Minerals, Metals & Materials Society): Pittsburgh, PA, USA, 2012; pp. 1095–1100.
23. Zohrabyan, D.; Milkereit, B.; Kessler, O.; Schick, C. Precipitation enthalpy during cooling of aluminum alloys obtained from calorimetric reheating experiments. *Thermochim. Acta* **2012**, *529*, 51–58. [CrossRef]
24. Fröck, H.; Milkereit, B.; Wiechmann, P.; Springer, A.; Sander, M.; Kessler, O.; Reich, M. Influence of Solution-Annealing Parameters on the Continuous Cooling Precipitation of Aluminum Alloy 6082. *Metals* **2018**, *8*, 265. [CrossRef]
25. Gannon, L. *Mesh Dependence of True Stress-Strain Curves in Finite Element Analysis of Steel Structures*; Defence R&D Canada-Atlantic: Dartmouth, NS, Canada, 2011.

© 2018 by the authors. Licensee MDPI, Basel, Switzerland. This article is an open access article distributed under the terms and conditions of the Creative Commons Attribution (CC BY) license (http://creativecommons.org/licenses/by/4.0/).

Article

Microstructure, Mechanical Properties, and Corrosion Resistance of Thermomechanically Processed AlZn6Mg0.8Zr Alloy

Aleksander Kowalski [1],*, Wojciech Ozgowicz [2], Wojciech Jurczak [3], Adam Grajcar [2], Sonia Boczkal [4] and Janusz Żelechowski [4]

1. Institute of Non-Ferrous Metals, 5 Sowińskiego Street, 44-100 Gliwice, Poland
2. Silesian University of Technology, Institute of Engineering Materials and Biomaterials, 18A Konarskiego Street, 44-100 Gliwice, Poland; wojciech.ozgowicz@tlen.pl (W.O.); adam.grajcar@polsl.pl (A.G.)
3. Polish Naval Academy, Faculty of Mechanical and Electrical Engineering, 69 Śmidowicza Street, 81-127 Gdynia, Poland; w.jurczak@amw.gdynia.pl
4. Institute of Non-Ferrous Metals, Light Metals Division, 19 Piłsudskiego Street, 32-050 Skawina, Poland; sboczkal@imn.skawina.pl (S.B.); jzelechowski@imn.skawina.pl (J.Ż)
* Correspondence: aleksander.kowalski@imn.gliwice.pl; Tel.: +48-32-2380-232

Received: 7 March 2018; Accepted: 4 April 2018; Published: 7 April 2018

Abstract: The paper presents results of the investigations on the effect of low-temperature thermomechanical treatment (LTTT) on the microstructure of AlZn6Mg0.8Zr alloy (7000 series) and its mechanical properties as well as electrochemical and stress corrosion resistance. For comparison of the LTTT effect, the alloy was subjected to conventional precipitation hardening. Comparative studies were conducted in the fields of metallographic examinations and static tensile tests. It was found that mechanical properties after the LTTT were better in comparison to after conventional heat treatment (CHT). The tested alloy after low-temperature thermomechanical treatment with increasing plastic deformation shows decreased electrochemical corrosion resistance during potentiodynamic tests. The alloy after low-temperature thermomechanical treatment with deformation degree in the range of 10 to 30% is characterized by a high resistance to stress corrosion specified by the level of P_{SCC} indices.

Keywords: aluminium alloy; 7003 alloy; corrosion resistance; thermomechanical treatment; TEM; X-ray diffraction

---

1. Introduction

The effect of microstructure and morphology of intermetallic phases on mechanical properties and corrosion resistance of Al–Zn–Mg wrought aluminium alloys is important due to the required properties of final products. The proper design of structural elements exposed to the aggressive impact of chloride ion medium requires comprehensive understanding of the relationships between microstructure, microsegregation of alloying elements, type and morphology of intermetallic precipitations, and the heat or thermomechanical treatment performed [1–5]. Refinement of microstructure and the size of precipitations, ensuring optimal strain hardening and precipitation strengthening of the 7000 series aluminium alloys as a result of thermomechanical processing, is an indispensable condition for technological development in the field of newly designed light constructions of machines and devices in many industrial fields [6–9].

Integration of heat treatment and plastic deformation operations of aluminium alloys significantly improves their impact efficiency on the microstructure, especially in the case of conventional techniques including cold rolling, solution heat treatment, and ageing. Advantageous effects of improved mechanical properties and corrosion resistance of Al–Zn–Mg alloys are obtained as a result of synergetic interactions of strain hardening and precipitation hardening [10–13]. Cold deformation of supersaturated solid solution reduces its thermodynamic stability and accelerates ageing effects. However, the influence of strain hardening on the ageing process is complicated, because it depends on conditions of supersaturation, deformation degree and ageing temperature, alloy type, and the type of precipitations occurring during ageing [14–17].

Nowadays, Al–Zn–Mg–Cu alloys and their various modifications with Zr, Sc, and Ti microadditions are most interesting among the high-strength aluminium alloys of the 7000 series. Deng et al. [11] investigated the possibility of counteracting recrystallization in the 7085 alloy by applying high-temperature thermomechanical treatment with subsequent precipitation hardening. In turn, Lee et al. [12] considered obtaining the fine-grained microstructure of the 7075 alloy using twin roll casting and thermomechanical treatment. Another modification of technological process of the 7075 alloy, which consisted of double thermomechanical processing leading to refinement of the microstructure, was proposed by El-Baradie et al. [13]. Zuo et al. [15–17] used two-stage hot rolling during thermomechanical treatment of the 7055 alloy for realization of the same purpose. In turn, Huo et al. [18,19] carried out continuous rolling during cooling from a hot rolling temperature for the 7075 alloy.

Adequate resistance in aggressive corrosive media is an important criterion in selecting Al–Zn–Mg alloys for responsible construction elements in various industries. Chemical composition becomes substantial in this aspect—primarily the total content of Zn and Mg, the amount of Cu, and microadditions of Zr, Sc, or Ti [20–22]. Moreover, microstructure development, ensuring high stress corrosion resistance while maintaining high mechanical properties, is also not to be neglected [23,24]. Wang et al. [25] investigated the stress corrosion resistance of an Al–Zn–Mg alloy with a small content of additions, subjected to two-stage precipitation hardening. Successively, comprehensive considerations on the effect of variant heat treatment of Al–Zn–Mg–Cu (7085) alloy on microstructure and mechanical properties, and, in particular, resistance to intercrystalline and stress corrosion, were the subject of research led by Peng et al. [26]. The effect of precipitation hardening parameters on microstructure of an Al–Zn–Mg–Sc–Zr alloy, which in turn determines stress corrosion resistance, was undertaken by Huang et al. [27]. Sun et al. [28] conducted research on the dependence of grain boundary structure and stress corrosion of an Al–Zn–Mg alloy without Cu and microadditions.

Available literature data mainly relate to Al–Zn–Mg alloys containing Cu. Difficulties in obtaining high-quality welded joints limit their application. Aluminium alloys of this series, with limited amounts of Cu, are superior in this aspect. There are, however, relatively few scientific studies on these alloys. The number of available publications dealing with the matter of thermomechanical treatment of the 7000 series alloys, and, in particular, its low-temperature variant with Zr microaddition, is also not high. Therefore, the development of knowledge, enabling optimal use of available methods of plastic working and heat treatment as well as their synergetic effect on improvement of functional properties of 7000 series aluminium alloys without the addition of Cu, is desirable. This is why studies of AlZn6Mg0.8Zr alloy were aimed at determining the impact of different reductions during low-temperature thermomechanical treatment on its microstructure, mechanical properties, and resistance to electrochemical and stress corrosion.

2. Materials and Methods

2.1. Materials and Heat Treatment

The material used was a metal sheet, cold-rolled from Al–Zn–Mg (7000 series) commercial aluminium alloy with the following chemical composition: 6.13 Zn, 0.74 Mg, 0.30 Mn, 0.20 Fe, 0.17 Cr, 0.12 Si, 0.08 Zr, 0.04 Cu, and the remainder Al. The investigated alloy was subjected to the following low-temperature thermomechanical treatment: supersaturation in water from the temperature of 500 °C after a holding time of 1 h; cold-rolled with a reduction of 10%, 20%, and 30%; strain-aged for

12 h at a temperature of 150 °C; then cooled down in the open air (Figure 1). For comparison purposes, the tested alloy was subjected to the following precipitation hardening: solution heat treatment (supersaturation) in water from the temperature of 500 °C after a holding time of 1 h; quench-aged for 12 h at a temperature of 150 °C; then cooled down in the open air.

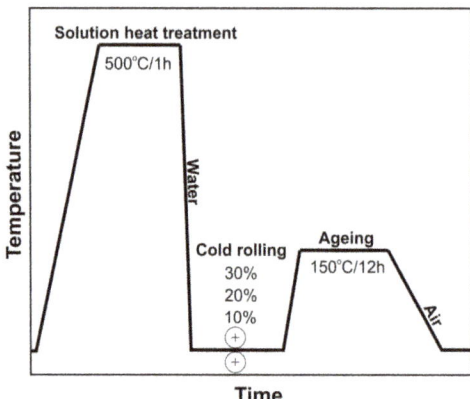

Figure 1. A schematic diagram of the low-temperature thermomechanical treatment (LTTT) of the AlZn6Mg0.8Zr alloy.

2.2. Tensile Test

Static tensile tests of the AlZn6Mg0.8Zr alloy were carried out at room temperature after low-temperature thermomechanical treatment and conventional precipitation hardening using an INSTRON 4505 universal testing machine at a traverse speed rate of 2 mm/min, according to the standard [29]. For examinations of mechanical properties, round samples (three for each state) with the following dimensions of the measuring part were used: Ø 10 mm and l_o = 60 mm. The presented results are an arithmetic average of three measurements.

2.3. Transmission Electron Microscopy

Transmission electron microscopy (TEM) tests were carried out using the thin foil technique. Discs of Ø 3 mm diameter, cut from the AlZn6Mg0.8Zr alloy after low-temperature thermomechanical treatment (LTTT) and conventional heat treatment (CHT), were ground on abrasive paper to a thickness of about 0.3 mm, and then electrolytically polished at temperature −20 °C at voltage 16.5 V in Struers A2 reagent using a TenuPol-5 device. The microstructure of the thin foils was observed using a TECNAI G2 (FEI) transmission electron microscope, at accelerating voltage of 200 kV and magnification up to 120000×. The phase identification procedure, based on electron diffraction, was performed with the aid of the ELDYF computer program [30].

2.4. X-ray Diffraction Examinations

X ray examinations were performed in the delivery state and after conventional heat treatment and LTTT operations. For this purpose, the D8 Advance (Bruker, Karlsruhe, Germany) X-ray diffractometer was used. The XRD pattern was obtained with the Bragg–Brentano method using X-ray radiation with energy of approximately 8.041 keV, which corresponds to with a wavelength of approximately 1.54184 Å. X-ray qualitative analysis was carried out in the range of θ angles from 10° to 40°. In order to obtain high-quality diffraction patterns of high accuracy and precision, the time of measurement cycle (step) was set to 100 s, which increases the recording time of one diffraction line ranging from a few to several hours at the step equal to 0.045° θ.

2.5. Electrochemical and Stress Corrosion Examinations

Electrochemical corrosion resistance tests were performed using the potentiodynamic method, in accordance with recommendations of the standard [31]. Corrosion resistance was analyzed on the basis of registered potentiodynamic curves of a logarithm dependence of current density versus potential value. Measurements were carried out in the 3.5% NaCl medium (pH = 7). The examinations were performed on three specimens for each state after LTTT and the arithmetic average of measurements is presented.

Stress corrosion resistance tests were executed at the PNA Gdynia stand at a constant tensile load of $\sigma_o = (0.8)R_{p0.2}$ (yield strength) in the medium of 3.5% NaCl aqueous solution. The corrosion test temperature was within the range of room temperature, and the NaCl aqueous solution was changed every two days. The shape of specimens is presented in Figure 2. The exposure time was equal to 1512 h, and three specimens for each state were used. The resistance of the tested alloy to stress corrosion was determined by comparing mechanical properties obtained in a static tensile test before and after corrosion exposure, using the following Formula (1):

$$P_{SCC} = (1 - \frac{P_{NaCl}}{P_{air}}) \cdot 100\% \tag{1}$$

where

P_{SCC}—stress corrosion susceptibility index for individual material properties;
P_{NaCl}—material property measured in corroding medium;
P_{air}—material property measured in air.

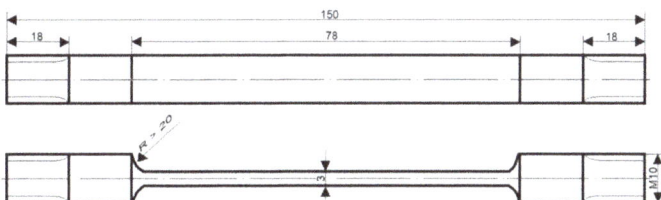

Figure 2. Shape and dimensions of stress corrosion test specimens.

3. Results and Discussion

3.1. Mechanical Properties and Microstructure

The AlZn6Mg0.8Zr alloy subjected to LTTT reveals a yield point from approximately 256 MPa to about 300 MPa and tensile strength from approximately 321 MPa to 347 MPa, depending on the degree of cold deformation after solution heat treatment in a range from 10 to 30% (Figure 3a). The obtained values of tensile strength are lower (approximately 250 MPa) in comparison with an alloy containing 8.38% Zn, 2.07% Mg, 2.31% Cu, and 0.13% Zr subjected to high-temperature thermomechanical treatment (HTTT) [15–17]. However, the strength difference decreases with the reduction of Cu and Zr contents. For example, an alloy containing 7.81% Zn, 1.62% Mg, 1.81% Cu, and 0.13% Zr after HTTT is characterized by 160 MPa higher tensile strength [11], while an alloy with a copper content limited to 0.1% Cu and 0.13% Zr subjected to more complex heat treatment shows a 30 MPa increase in tensile strength [8]. It should be noted that these values depend significantly on processing conditions. The value of the $R_{p0.2}/R_m$ ratio varies along with the deformation degree from 0.8 to 0.87. Elongation of the examined alloy takes similar values, whereas reduction of area decreases by approximately 9% for a given deformation range (Figure 3b). Obtained values of mechanical properties after thermomechanical treatment are higher when compared to the values obtained after CHT (Table 1). A significant rise in yield point has been noted. Thermomechanical treatment causes a slight decrease of elongation. Nevertheless, in comparison to the alloy subjected to the CHT, the tested alloy subjected to the LTTT is characterized by higher strength and reduction of area in the entire range of deformation.

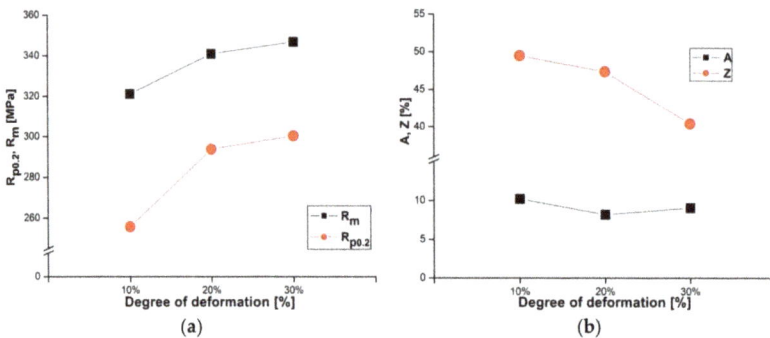

Figure 3. Influence of deformation degree on (**a**) strength and (**b**) plastic properties of the investigated alloy after LTTT.

Table 1. Mechanical properties of the investigated alloy after the LTTT and conventional heat treatment (CHT).

	Treatment Parameters			Mechanical Properties			
			LTTT				
Temperature of Solution Heat Treatment (°C)	Temperature of Ageing (°C)	Time of Ageing (h)	Degree of Deformation (%)	$\overline{R_{p0.2}}$ (MPa)	$\overline{R_m}$ (MPa)	$\overline{A}$(%)	$\overline{Z}$(%)
500	150	12	10%	256 ± 4	321 ± 2	10.2 ± 0.4	50 ± 5
			20%	294 ± 2	341 ± 2	8.2 ± 1.1	47 ± 5
			30%	301 ± 2	347 ± 2	9.1 ± 0.9	40 ± 4
			CHT				
500	150	12	—	230 ± 3	310 ± 2	16.8 ± 0.5	34 ± 4
			R_m—tensile strength; A—total elongation; Z—reduction in area.				

The $R_{p0.2}/R_m$ ratio should also be taken into consideration. The AlZn6Mg0.8Zr alloy subjected to LTTT is characterized by a higher $R_{p0.2}/R_m$ ratio (about 0.87), in comparison to the CHT carried out (about 0.74) under comparable processing conditions. Generally, higher values of this ratio are observed for aluminium alloys with a smaller fraction of recrystallized microstructure, higher strengthening from the substructure, and lower elongation [11], which corresponds well with obtained results.

Observations of thin foils using transmission electron microscopy (TEM) revealed after the CHT the substructure of the α solution matrix composed of subgrains ranging from 1 μm to about 2 μm, inside which fine, oval, and nonuniformly distributed precipitates were found (Figure 4a). The presence of grain boundaries with particles arranged along these boundaries (Figure 4b), often in a continuous manner (Figure 4c), was also revealed. These precipitates are also distributed uniformly inside grains or they are stochastic. It has also been found that there are precipitation-free zones (PFZ) in the vicinity of grain boundaries (Figure 4c). This microstructure is characteristic for Al–Zn–Mg alloys subjected to heat treatment [25–27]. Deschamps et al. [9] found that the critical factor shaping the microstructure of Al–Zn–Mg–Zr alloys, in addition to degradation and morphology of the Al_3Zr phase, is also the volume fraction of GP zones, because they are privileged nucleation areas of the η′ phase. Electron diffractions revealed reflexes coming from the α solution matrix (Al) and $MgZn_2$ phase (Figure 5).

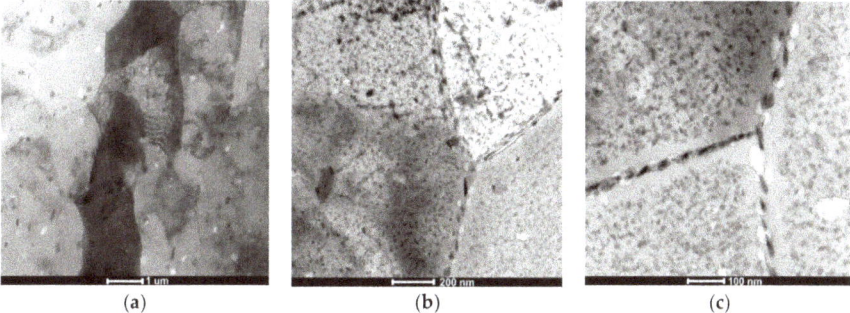

Figure 4. The AlZn6Mg0.8Zr alloy after CHT: (**a**) substructure; (**b**) precipitations inside grains and along grains' boundaries; (**c**) precipitation-free zones (PFZ) near to grains' boundaries with continuously distributed precipitates.

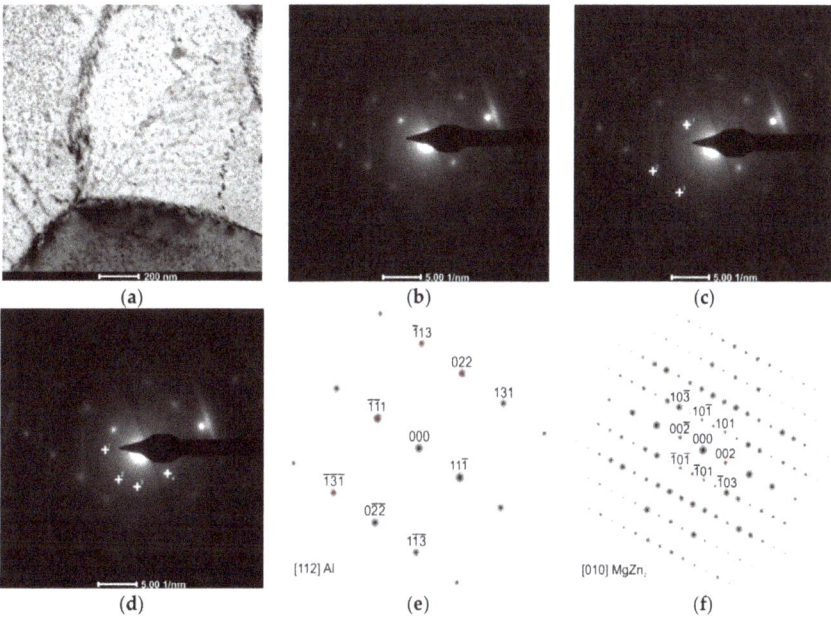

Figure 5. The AlZn6Mg0.8Zr alloy after CHT: (**a**) substructure of investigated alloy with stochastic distribution of precipitations inside subgrains and along their boundaries; (**b**) electron diffraction; (**c**) marked reflections of α solution matrix; (**d**) marked reflections of MgZn$_2$ particle; (**e**) solution of α phase matrix diffraction and (**f**) solution of MgZn$_2$ phase diffraction.

The alloy subjected to the LTTT with 10% reduction is characterized by the microstructure composed of subgrains of the α solution with size in the range of 0.5–1 μm (Figure 6a). Fine, oval precipitates are usually distributed along grain boundaries (Figure 6b). They do not form continuous nets. Along these boundaries, a narrow precipitation-free zone can be observed—less distinct than in the case of the CHT (Figure 6c). Diffraction analysis of these particles allowed the identification of η–MgZn$_2$ particles (Figure 7).

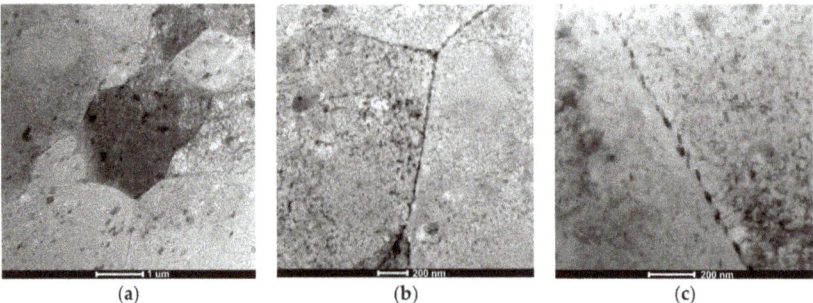

Figure 6. The AlZn6Mg0.8Zr alloy after LTTT with 10% deformation: (**a**) α solution matrix with nonuniform distribution of precipitates; (**b**) fine chain-arranged particles along grain boundaries; (**c**) PFZ along grain boundary with precipitates.

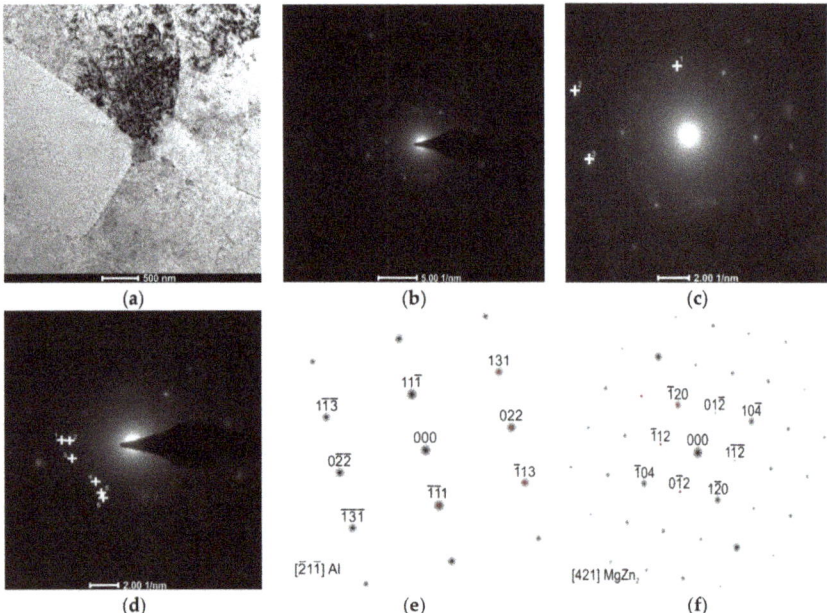

Figure 7. The AlZn6Mg0.8Zr alloy after LTTT with 10% deformation: (**a**) stochastically distributed precipitates in the substructure; (**b**) electron diffraction; (**c**) marked reflections of α solution matrix; (**d**) marked reflections of $MgZn_2$; (**e**) solution of α phase matrix diffraction and (**f**) solution of $MgZn_2$ phase diffraction.

Observations of thin foils of the alloy subjected to LTTT with 30% reduction revealed the presence of a cellular dislocation microstructure with a high density of dislocations (Figure 8a). This effect is caused by inhibition of the mobile dislocation movement by η phase precipitates and Al_3Zr [15,18]. Numerous nonuniformly distributed precipitates were revealed in the matrix of the α phase. They are elongated, oval, and concentrated both inside grains and at their boundaries. It was found that particles are distributed discontinuously along grain and subgrain boundaries. Decay of precipitation-free zones in the vicinity of these boundaries was also observed (Figure 8b). Precipitates at the boundaries of grains reveal a characteristic arrangement of parallel lines on the surface with relatively regular spacing, which in the literature has been called the "striation effect", characteristic for nanometric-sized precipitates [32]. This effect is metallographically similar to the patterns of dense stacking faults (Figure 8c). Based on electron diffraction, the presence of the MgZn$_2$ phase was confirmed in the substructure of the alloy (Figure 9).

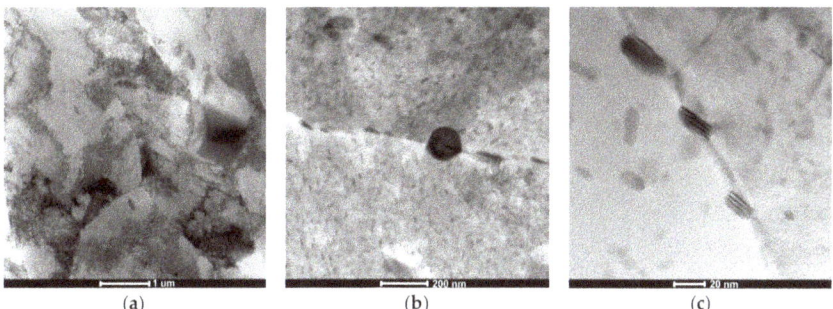

Figure 8. The AlZn6Mg0.8Zr alloy after LTTT with 30% deformation: (**a**) cellular dislocation structure; (**b**) chain-arranged system of precipitates along subgrain boundary; (**c**) morphology of MgZn$_2$ particle.

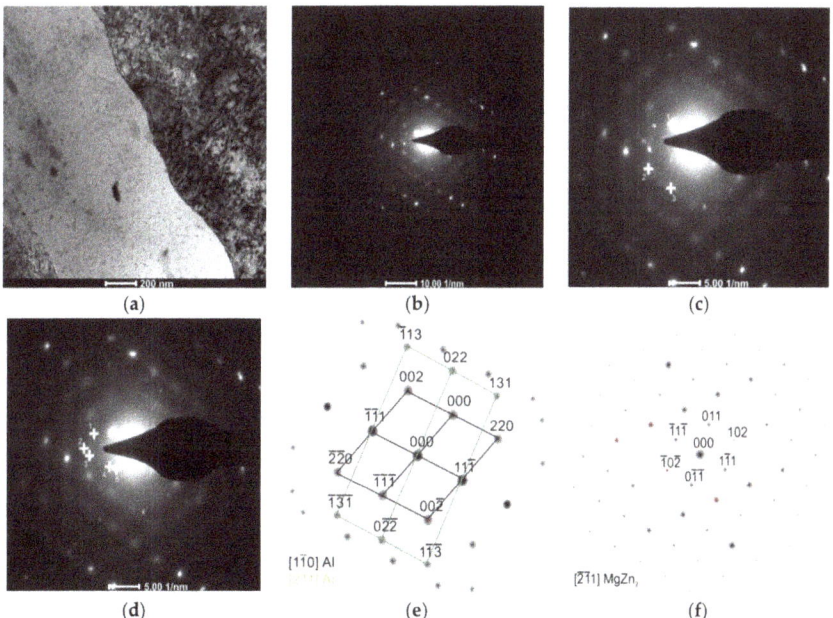

Figure 9. The AlZn6Mg0.8Zr alloy after LTTT with 30% deformation: (**a**) irregular subgrain boundaries; (**b**) electron diffraction; (**c**) marked reflections of α solution matrix; (**d**) marked reflections of MgZn$_2$ precipitate; (**e**) solution of α phase matrix diffraction and (**f**) solution of MgZn$_2$ phase diffraction.

Microstructure observations of the alloy after the LTTT indicate a substantial role of plastic deformation, which influences formation of a substructure, primarily in the aspect of secondary phase precipitates. This is the result of the formation of a higher density of dislocations in the supersaturated α solution, which not only accelerate disintegration of this solution, but act as preferred nucleation areas of precipitation even before accelerated ageing during deformation, in the process of dynamic strain ageing and as an effect of hereditary impact—so-called static strain ageing—while soaking the alloy at the ageing temperature [15]. This is due to an increased diffusion rate in these processes and the presence of numerous nuclei in the deformed α solution, as well as faster coagulation of these phases. Similar results of the impact of plastic deformation on the microstructure of Al–Zn–Mg-type alloys have been presented in the literature [16–19]. The discussed changes in microstructure significantly affect obtained mechanical properties and, in particular, the dependencies between the level of strength and plastic properties [33,34]. On the one hand, attention is paid to the impact of the increased amount of precipitates formed in the LTTT process on the level of alloy strengthening. On the other hand, it influences a morphology, especially the degree of coagulation and distribution of formed particles. They affect a higher level of reduction in area after the LTTT compared with after conventional heat treatment.

3.2. Precipitation Behavior

Qualitative phase analysis of the AlZn6Mg0.8Zr alloy after the ageing stage in the CHT process revealed diffraction reflexes from the α matrix (Al) and Al_3Zr and $MgZn_2$ phases (Figure 10a). The same phases were identified in the alloy subjected to 30% cold deformation (Figure 10b). Moreover, the analysis of the intensity of recorded diffraction lines of the α solution matrix (Al), especially in case of the LTTT, indicates the lack of their compatibility with standard data.

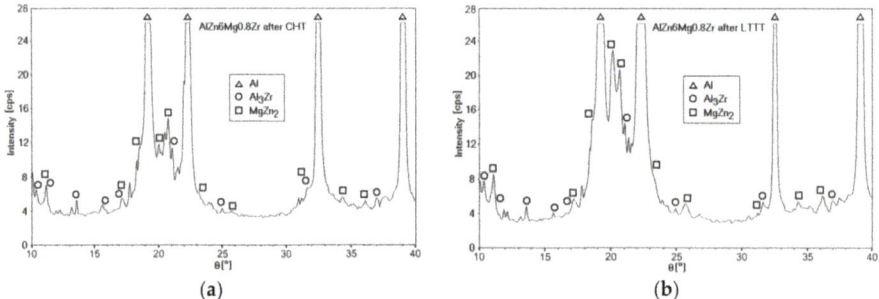

Figure 10. XRD patterns of the AlZn6Mg0.8Zr alloy after: (a) CHT (T_s = 500 °C; T_a = 150 °C; $τ_a$ = 12h); (b) LTTT (T_s = 500 °C; 30%; T_a = 150 °C; $τ_a$ = 12h).

The $MgZn_2$ phase is most often identified by X-ray phase analysis [15,19,21,25]. Extending the time of the measurement step to 100s allowed the obtaining of more accurate diffraction patterns, which enabled the identification of Al_3Zr dispersoids. The presence of fine-dispersive Al_3Zr phases in alloys of the 7000 series, which are formed already during cooling of the alloy after solution heat treatment, is important because they can become privileged areas of heterogeneous nucleation of the strengthening η–$MgZn_2$ phase [7]. The occurrence of Al_3Zr particles in combination with the high dislocation density, in the case of the LTTT processing, causes formation of more nucleation sites for the η phase and, thus, a higher strength of alloys containing Zr [9,11].

3.3. Corrosion Resistance

The AlZn6Mg0.8Zr alloy subjected to the LTTT with 10% and 20% reductions has a comparable corrosion potential (approximately −830 mV) (Figure 11). An increase in degree of deformation to 30% causes a decrease in E_{cor} by approximately 35 mV toward more active values. The value of the I_{cor} index increases and R_p decreases along with the increase of the deformation (Table 2). It was found that increase of the deformation amount leads to a decrease of electrochemical corrosion resistance. This is due to the formation of larger fraction of precipitates. They interact with the matrix and undergo anodic stripping, leading to the deterioration of electrochemical pitting corrosion factors [21,22,28].

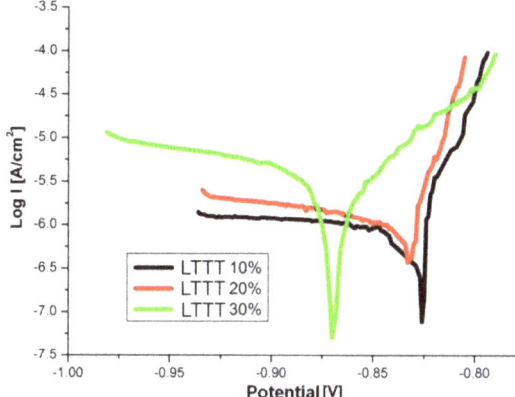

Figure 11. Polarization curves of the alloy after the LTTT with plastic deformation of 10%, 20%, and 30%.

Table 2. Electrochemical parameters obtained from cyclic polarization measurements.

Degree of Deformation (%)	E_{cor} (mV)	I_{cor} (µA/cm^2)	R_p (kΩ)
10	−830 ± 1	0.30 ± 0.2	17.4 ± 0.5
20	−833 ± 3	1.32 ± 0.3	6.2 ± 0.2
30	−867 ± 4	11.17 ± 0.7	5.4 ± 0.9

The alloy subjected to the LTTT is characterized by a slight decrease in mechanical properties in a static tensile test after loading the samples under stress corrosion conditions (Table 3). It was found that the decrease of mechanical factors after stress and corrosion exposures of samples is reduced when the degree of deformation rises. High values of stress corrosion susceptibility indices of the tested alloy (P_{SCC}) allow us to state that the AlZn6Mg0.8Zr alloy, subjected to the LTTT, is not very susceptible to this type of corrosion (Table 4). Similar conclusions for an Al–Zn–Mg–Zr alloy containing a small amount of Cu, subjected to various conventional heat treatments, were obtained by Xiao et al. [8]. In their case, the decrease in mechanical properties was more pronounced. It was also observed that P_{SCC} indices, for certain strength properties, are smaller than plastic properties indices (Z_{SCC} in particular). This indicates that strength properties are less susceptible to the effects of stress corrosion conditions. A decrease in all P_{SCC} indexes was also noted along with an increase in the deformation degree, which reflects improved stress corrosion resistance of the alloy along with its strengthening.

Cold deformation after solution heat treatment results in formation of a dislocation network, providing privileged areas for heterogeneous nucleation of the η phase. Moreover, new diffusion paths are created, which affect the precipitation sequence of the strengthening phases. As a consequence, the amount of η phase formed in the deformed alloy is larger than the portion of η' phase, which positively influences the resistance under stress corrosion conditions [24]. Moreover, ageing activated diffusion of alloying elements from inside to grain boundaries of cold-deformed alloys. This results in growth and coagulation of precipitates at the boundaries and their distribution in large intervals between them. The precipitation-free zone is also reduced [25–27,35].

Table 3. Mechanical properties of AlZnMg0.8Zr alloy, subjected to LTTT, obtained after and before stress corrosion tests.

Degree of Deformation (%)	$\overline{R}_{p0.2}$ (MPa)	$\overline{R}_m$ (MPa)	$\overline{A}$ (%)	$\overline{Z}$ (%)
Before Corrosion				
10	256 ± 4	321 ± 2	10.2 ± 0.4	50 ± 5
20	294 ± 2	341 ± 2	8.2 ± 1.1	47 ± 5
30	301 ± 2	347 ± 2	9.1 ± 0.9	40 ± 4
After Corrosion				
10	249 ± 2	316 ± 2	9.8 ± 1.3	46 ± 3
20	289 ± 3	337 ± 3	8.0 ± 0.7	45 ± 5
30	300 ± 4	346 ± 4	8.9 ± 0.5	39 ± 5

Table 4. P_{SCC} indices for the investigated alloy after LTTT.

Degree of Deformation (%)	P_{SCC} Index (%)			
	$R_{p0.2SCC}$	R_{mSCC}	A_{SCC}	Z_{SCC}
10	2.7	1.7	3.9	7.9
20	1.7	1.1	2.4	5.5
30	0.2	0.2	2.2	4.7

4. Conclusions

Studies of mechanical properties and corrosion resistance of the 7003 series Al–Zn–Mg-type alloy and metallographic analyses lead to the following conclusions:

- Low-temperature thermomechanical treatment with 30% reduction after solution heat treatment in water from the temperature of 500 °C and with ageing at the temperature of 150 °C ensures higher mechanical properties of the alloy in comparison to CHT.
- AlZn6Mg0.8Zr alloy reveals a microstructure consisting of the α solution matrix and fine-dispersive particles of morphologically differentiated intermetallic phases of the Al_3Zr and η–$MgZn_2$ types. Increase of cold deformation results in obtaining a smaller precipitation-free zone after ageing and formation of a greater portion of strengthening phases in the vicinity of grain boundaries. This affects the increase in the distance between precipitates located at grain boundaries.
- Electrochemical corrosion resistance of the AlZn6Mg0.8Zr alloy in the 3.5% NaCl medium decreases along with increasing plastic deformation degree, whereas the opposite behavior occurs in the case of stress corrosion resistance, which is improved. This is related to complex microstructural phenomena, which must be studied in more detail.

Author Contributions: A.K. and W.O. conceived and designed the experiments; A.K. performed tensile tests, analyzed the data and wrote the paper; W.O. supervised the work; W.J. performed stress corrosion examinations; A.G. analyzed the data and reviewed the paper; S.B. performed TEM experiments and analyzed the results; J.Ż. performed X-ray diffraction examinations and analyzed the results; All authors discussed the paper.

Conflicts of Interest: The authors declare no conflict of interest.

References

1. Hyde, K.B.; Norman, A.F.; Prangnell, P.B. The effect of cooling rate on the morphology of primary Al_3Sc inter-metallic particles in Al-Sc alloys. *Acta Mater.* **2001**, *49*, 1327–1337. [CrossRef]
2. Davydov, V.G.; Rostova, T.D.; Zakharov, V.V.; Filatov, Yu.A.; Yelagin, V.I. Scientific principles of making an alloying addition of scandium to aluminium alloys. *Mater. Sci. Eng. A* **2000**, *280*, 30–36. [CrossRef]
3. Chen, G.; Zhang, Y.; Du, Z. Mechanical behavior of Al-Zn-Mg-Cu alloy under tension in semi-solid state. *Trans. Nonferrous Met. Soc. China* **2016**, *26*, 643–648. [CrossRef]

4. Khalid Rafi, H.; Janaki Ram, G.D.; Phanikumar, G.; Prasad Rao, K. Microstructure and tensile properties of friction welded aluminium alloy AA7075-T6. *Mater. Design* **2010**, *31*, 2375–2380. [CrossRef]
5. Fuller, C.B.; Krause, A.R.; Dunand, D.C.; Seidman, D.N. Microstructure and mechanical properties of a 5754 aluminum alloy modified by Sc and Zr additions. *Mater. Sci. Eng. A* **2002**, *338*, 8–16. [CrossRef]
6. Xiang, H.; Pan, Q.L.; Yu, X.H.; Huang, X.; Sun, X.; Wang, X.D.; Li, M.J. Superplasticity behaviors of Al-Zn-Mg-Zr cold-rolled alloy sheet with minor Sc addition. *Mater. Sci. Eng. A* **2016**, *676*, 128–137. [CrossRef]
7. Lü, X.; Guo, E.; Rometsch, P.; Wang, L. Effect of on-step and two-step homogenization treatments on distribution of Al$_3$Zr dispersoids in commercial AA7150 aluminium alloy. *Trans. Nonferrous Met. Soc. China* **2012**, *22*, 2645–2651. [CrossRef]
8. Xiao, T.; Deng, Y.; Ye, L.; Lin, H.; Shan, Ch.; Qian, P. Effect of three-stage homogenization on mechanical properties and stress corrosion cracking of Al-Zn-Mg-Zr alloys. *Mater. Sci. Eng. A* **2016**, *675*, 280–288. [CrossRef]
9. Deschamps, A.; Bréchet, Y. Influence of quench and heating rates on the ageing response of an Al–Zn–Mg–(Zr) alloy. *Mater. Sci. Eng. A* **1998**, *251*, 200–207. [CrossRef]
10. Kowalski, A.; Ozgowicz, W.; Grajcar, A.; Lech-Grega, M.; Kurek, A. Microstructure and Fatigue Properties of AlZn6Mg0.8Zr Alloy Subjected to Low-Temperature Thermomechanical Processing. *Metals*, **2017**, *7*, 488. [CrossRef]
11. Deng, Y.; Zhang, Y.; Wan, L.; Zhang, X. Effect of thermomechanical processing on production of Al-Zn-Mg-Cu alloy plate. *Mat. Sci. Eng. A* **2012**, *554*, 33–40. [CrossRef]
12. Lee, Y.; Kim, W.; Jo, D.; Lim, Ch.; Kim, H. Recrystallization behavior of cold rolled Al-Zn-Mg-Cu fabricated by twin roll casting. *Trans. Nonferrous Met. Soc. China* **2014**, *24*, 2226–2231. [CrossRef]
13. El-Baradie, Z.M.; El-Sayed, M. Effect of double thermomechanical treatments on the properties of 7075 Al alloy. *J. Mater. Process. Tech.* **1996**, *62*, 76–80. [CrossRef]
14. Ozgowicz, W.; Kalinowska, E.; Kowalski, A.; Gołombek, K. The structure and mechanical properties of Al-Mg-Mn alloys shaped in the process of thermomechnical treatment. *J. Achiev. Mater. Manuf. Eng.* **2011**, *45*, 148–156.
15. Zuo, J.; Hou, L.; Shi, J.; Cui, H.; Zhuang, L.; Zhang, J. The mechanism of grain refinement and plasticity enhancement by an improved thermomechanical treatment of 7055 Al alloy. *Mat. Sci. Eng. A* **2017**, *702*, 42–52. [CrossRef]
16. Zuo, J.; Hou, L.; Shi, J.; Cui, H.; Zhuang, L.; Zhang, J. Effect of deformation induced precipitation on grain refinement and improvement of mechanical properties AA 7055 aluminum alloy. *Mater. Charact.* **2017**, *130*, 123–134. [CrossRef]
17. Zuo, J.; Hou, L.; Shi, J.; Cui, H.; Zhuang, L. Enhanced plasticity and corrosion resistance of high strength Al-Zn-Mg-Cu alloy processed by an improved thermomechanical processing. *J. Alloy Compd.* **2017**, *716*, 220–230. [CrossRef]
18. Huo, W.T.; Shi, J.T.; Hou, L.G.; Zhang, J.S. An improved thermo-mechanical treatment of high-strength Al-Zn-Mg-Cu alloy for effective grain refinement and ductility modification. *J. Mater. Process. Tech.* **2017**, *239*, 303–3014. [CrossRef]
19. Huo, W.; Hou, L.; Cui, H.; Zhuang, L.; Zhang, J. Fine-grained AA 7075 processed by different thermo-mechanical processings. *Mat. Sci. Eng. A* **2014**, *618*, 244–253. [CrossRef]
20. Segal, V.M. New hot thermo-mechanical processing of heat treatable aluminum alloys. *J. Mater. Process. Tech.* **2016**, *231*, 50–57. [CrossRef]
21. Navaser, M.; Atapour, M. Effect of friction stir processing on pitting corrosion and intergranular attack of 7075 aluminum alloy. *J. Mater. Sci. Technol.* **2017**, *33*, 155–165. [CrossRef]
22. Li, S.; Guo, D.; Dong, H. Effect of flame rectification on corrosion property of Al-Zn-Mg alloy. *Trans. Nonferrous Met. Soc. China* **2017**, *27*, 250–257. [CrossRef]
23. Lu, X.; Han, X.; Du, Z.; Wang, G.; Lu, L.; Lei, J.; Zhou, T. Effect of microstructure on exfoliation corrosion resistance in an Al-Zn-Mg alloy. *Mater. Charact.* **2018**, *135*, 167–174. [CrossRef]
24. Umamaheshwer Rao, A.C.; Vasu, V.; Govindaraju, M.; Sai Srinadh, K.V. Stress corrosion cracking behavior of 7xxx aluminum alloys: A literature review. *Trans. Nonferrous Met. Soc. China* **2016**, *26*, 1447–1471.
25. Wang, Y.L.; Jiang, H.C.; Li, Z.M.; Yan, D.S.; Zhang, D.; Rong, L.J. Two-stage double peaks ageing and its effect on stress corrosion cracking susceptibility of Al-Zn-Mg alloy. *J. Mater. Sci. Technol.*. (in press). [CrossRef]

26. Peng, X.; Guo, Q.; Liang, X.; Deng, Y.; Gu, Y.; Xu, G.; Yin, Z. Mechanical properties, corrosion behavior and microstructures of a non-isothermal ageing treated Al-Zn-Mg-Cu alloy. *Mat. Sci. Eng. A* **2017**, *688*, 146–154. [CrossRef]
27. Huang, X.; Pan, Q.; Li, B.; Liu, Z.; Huang, Z.; Yin, Z. Microstructure, mechanical properties and stress corrosion cracking of Al-Zn-Mg-Zr alloy sheet with trace amount of Sc. *J. Alloy Compd.* **2015**, *650*, 805–820. [CrossRef]
28. Sun, X.Y.; Zhang, B.; Lin, H.Q.; Zhou, Y.; Sun, L.; Wang, J.Q.; Han, E.-H.; Ke, W. Correlations between stress corrosion cracking susceptibility and grain boundary microstructures for an Al-Zn-Mg alloy. *Corros. Sci.* **2013**, *77*, 103–112. [CrossRef]
29. ISO 6892-1:2016. *Metallic Materials—Tensile Testing—Part 1: Method of Test at Room Temperature*; Chinese Code: Beijing, China, 2010.
30. Gigla, M.; Pączkowski, P. The computer aided analysis of electron diffraction patterns. *Arch. Mater. Sci.* **2006**, *1*, 49–69.
31. ISO 17475:2015. *Corrosion of Metals and Alloys—Electrochemical Test Methods—Guidelines for Conducting Potentiostatic and Potentiodynamic Polarization Measurements*; Chinese Code: Beijing, China, 2010.
32. Ozgowicz, W. *Physicochemical, Structural and Mechanical Factors of Intergranular Brittleness of the α Bronzes at Elevated Temperature*; Scientific Notebooks of Silesian University of Technology: Gliwice, Poland, 2004.
33. Han, N.; Zhang, X.; Liu, S.; Ke, B.; Xin, X. Effects of pre-stretching and ageing on the strength and fracture toughness of aluminum alloy 7050. *Mat. Sci. Eng. A* **2011**, *528*, 3714–3721. [CrossRef]
34. Srivatsan, T.S.; Sriram, S.; Veeraraghavan, D.; Vasudevan, V.K. Microstructure, tensile deformation and fracture behaviour of aluminium alloy 7055. *J. Mater. Sci.* **1997**, *32*, 2883–2894. [CrossRef]
35. Wang, D.; Ma, Z.Y.; Gao, Z.M. Efect of severe cold rolling on tensile properties and stress corrosion cracking of 7050 aluminum alloy. *Mater. Chem. Phys.* **2009**, *17*, 228–233. [CrossRef]

© 2018 by the authors. Licensee MDPI, Basel, Switzerland. This article is an open access article distributed under the terms and conditions of the Creative Commons Attribution (CC BY) license (http://creativecommons.org/licenses/by/4.0/).

Article

Revealing the Effect of Local Connectivity of Rigid Phases during Deformation at High Temperature of Cast AlSi12Cu4Ni(2,3)Mg Alloys

Katrin Bugelnig [1,2,*], Holger Germann [3], Thomas Steffens [3], Federico Sket [4], Jérôme Adrien [5], Eric Maire [5], Elodie Boller [6] and Guillermo Requena [2,7]

1. Institute of Materials Science and Technology, Technical University of Vienna, 13/308 Karlsplatz, A-1040 Vienna, Austria
2. German Aerospace Centre, Linder Höhe, 51147 Cologne, Germany; guillermo.requena@dlr.de
3. KS Kolbenschmidt GmbH, Karl-Schmidt-Straße, 74172 Neckarsulm, Germany; Holger.Germann@de.rheinmetall.com (H.G.); Thomas.Steffens@de.rheinmetall.com (T.S.)
4. IMDEA Materials Institute, C/Eric Kandel 2, 28906 Getafe, Spain; federico.sket@imdea.org
5. Laboratoire MATEIS, UMR5510 CNRS, INSA Lyon, Université de Lyon, 69621 Villeurbanne, France; jerome.adrien@insa-lyon.fr (J.A.); eric.maire@insa-lyon.fr (E.M.)
6. ESRF—The European Synchrotron, CS40220 Grenoble CEDEX 9, France; boller@esrf.fr
7. Metallic Structures and Materials Systems for Aerospace Engineering, RWTH Aachen University, 52062 Aachen, Germany
* Correspondence: Katrin.Bugelnig@dlr.de or katrin.bugelnig@tuwien.ac.at; Tel.: +49-2203-601-2230

Received: 6 July 2018; Accepted: 23 July 2018; Published: 27 July 2018

Abstract: The 3D microstructure and its effect on damage formation and accumulation during tensile deformation at 300 °C for cast, near eutectic AlSi12Cu4Ni2Mg and AlSi12Cu4Ni3Mg alloys has been investigated using in-situ synchrotron micro-tomography, complemented by conventional 2D characterization methods. An increase of Ni from 2 to 3 wt.% leads to a higher *local connectivity*, quantified by the Euler number χ, at constant *global interconnectivity* of rigid 3D networks formed by primary and eutectic Si and intermetallics owing to the formation of the plate-like Al-Ni-Cu-rich δ-phase. Damage initiates as micro-cracks through primary Si particles agglomerated in clusters and as voids at matrix/rigid phase interfaces. Coalescence of voids leads to final fracture with the main crack propagating along damaged rigid particles as well as through the matrix. The lower local connectivity of the rigid 3D network in the alloy with 2 wt.% Ni permits localized plastification of the matrix and helps accommodating more damage resulting in an increase of ductility with respect to AlSi12Cu4Ni3Mg. A simple load partition approach that considers the evolution of local connectivity of rigid networks as a function of strain is proposed based on in-situ experimental data.

Keywords: Cast Al-Si alloys; 3D characterization; synchrotron tomography; in-situ tensile deformation; 3D microstructure; damage; strength; connectivity

1. Introduction

The strength of cast Al-Si piston alloys is given by the strength of the age-hardenable α-Al matrix and the load carrying capability of interconnected 3D hybrid networks formed by Si and various intermetallic phases [1–5]. The alloying elements Cu and Mg play a decisive role on precipitation strengthening of the matrix, while, together with Ni, they define the interconnectivity of the 3D networks [6–9]. These rigid networks exhibit high thermal stability and provide strength retention up to about 300 °C even after overaging of the matrix [10,11].

In recent years, the influence of the interconnectivity of the 3D networks on strength at ambient and elevated temperatures has been made evident by several investigations [1,11–15]. It is now well established that these networks have complex morphologies and it is therefore necessary to describe them considering their connectivity as a whole structure, termed here global interconnectivity [2], as well as connections within the network, that is, their local connectivity [16,17]. Thus, the ambient and elevated temperature strengths of Al-Si piston alloys are improved in the presence of ~20 vol.% of highly interconnected (>95% global interconnectivity) networks consisting of eutectic Si and Ni-, Cu-, Fe-rich intermetallics owing to a transfer of mechanical load from the weaker matrix to these rigid networks [1,11–14,18,19]. It is also well known now that Ni plays a decisive role in terms of enhancement of strength of these alloys, especially at high temperatures up to 300 °C [6,18–21]. An addition of at least 1 wt.% of Ni to a cast AlSi12 alloy resulted in the formation of 3D networks with a high contiguity between Si and aluminides and this reduced the spheroidisation of eutectic Si during solution treatment [11,12]. Moreover, an increase of Ni content from 1 to 2 wt.% in an AlSi10Cu5NiX was observed to result in an increase of room temperature strength by ~15% after 4 h solution treatment at 500 °C [1,15]. The reason for this is that, although in the as cast condition similar degrees of global interconnectivity of the rigid phases (Si + aluminides) are obtained, the addition of more than 1 wt.% Ni is necessary to avoid the partial disintegration of the 3D network during solution treatment.

It is however insufficient to consider only the global interconnectivity of these rigid 3D networks to understand the mechanical behavior of Al-Si piston alloys. We have recently shown that the room temperature tensile strength of an AlSi12Cu4Ni2 alloy [16] decreases after 4 h solution treatment at 500 °C although the strength of the α-Al matrix and the global interconnectivity of rigid networks remain constant. A closer look using synchrotron micro-tomography with μm resolution revealed that changes in the local connectivity of the networks occur during solution treatment as a result of partial dissolution of Al_2Cu aluminides and preliminary states of Al_2Cu (e.g., θ′, θ″) as well as slight spheroidisation and fragmentation of eutectic and primary Si particles. These changes were quantified using the topological parameter Euler number χ [17,22–24] and it was suggested that a decrease of local connectivity owing to the loss of connecting branches within the 3D network provoked a decrease in the load bearing capability of the globally fully interconnected 3D networks. The Euler number χ has been used to study theoretically the effect of local connectivity on strength of a periodic 2D microstructure by Silva et al. [24]. For this purpose, they randomly removed connections in their model material and found that a loss of 10% of connections resulted in a decrease of strength of about 35%. Moreover, Kruglova et al. [17] theoretically correlated an increase in strength of an AlSi7 alloy to a more negative Euler number of the 3D network formed by eutectic Si, that is, higher local connectivity of the load bearing phase. Thus, our experimental results at room temperature [16] and these theoretical studies clearly indicate that local connectivity should be taken into account to understand the mechanical behavior of these alloys. However, it must also be considered that the 3D rigid networks undergo damage during deformation and, therefore, both their global and local interconnectivities may gradually change. This means that the load carrying capability of the networks changes as well gradually as deformation advances.

In the current work, we present an experimental approach to quantify the evolution of local connectivity of rigid 3D networks as a function of damage during tensile deformation at 300 °C of AlSi12Cu4NixMg piston alloys (x = 2–3) using in-situ synchrotron X-ray computed tomography. This is to the best of our knowledge the first study that comprises quantification of the evolution of interconnectivity of 3D networks in Al-Si alloys experimentally and the introduction of a simple analytical load partition model that considers local connectivity changes as a function of strain.

2. Materials and Methods

2.1. Materials

Two cast near-eutectic Al-Si piston alloys with the chemical compositions given in Table 1 were investigated.

Table 1. Chemical compositions of the investigated alloys (wt.%).

Alloy	Al	Si	Cu	Ni	Mg
AlSi12Cu4Ni2Mg	bal.	12.5	4	2	1
AlSi12Cu4Ni3Mg	bal.	13.1	4	3	1

Piston raw parts produced by gravity die casting were manufactured by KS Kolbenschmidt GmbH, Neckarsulm, Germany. All the samples studied in this work were taken from the bowl rim area of these pistons (see Figure 1). Both alloys were subjected to ageing at 230 °C for 5 h followed by air cooling.

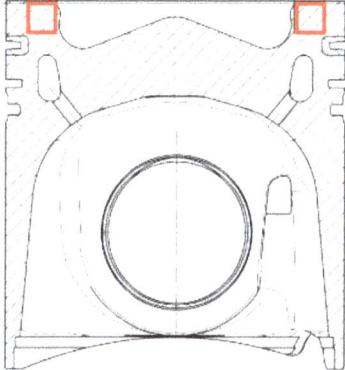

Figure 1. Illustration of the cross-section of a piston with the location of sample extraction at the bowl rim area indicated with red squares.

2.2. Methods

Light optical microscopy (LOM) and scanning electron microscopy (SEM) were complemented by synchrotron X-ray computed tomography (sXCT). sXCT can provide the necessary phase contrast to reveal simultaneously the α-Al matrix as well as eutectic and primary Si. This is not possible by laboratory XCT, owing to the very similar X-ray attenuations of these phases [2]. For a qualitative investigation of the interconnectivity of rigid phases, specimens were chemically etched using a H_2O + HCl solution with a ratio of 60:40 for approximately 45 min to ensure a slow and gentle dissolution of the α-Al matrix without damage of the rigid phases or distortion of the hybrid network. Thereafter, scanning electron microscopy of the etched specimens was conducted and complemented with energy-dispersive X-ray spectroscopy (EDX).

The hardness of the alloys and of the α-Al matrix was determined by Brinell hardness HB (1/10) and nano-indentation, respectively. To obtain statistically relevant values, at least 5 Brinell hardness measurements were conducted on the sample surface of each specimen, while for the nano-hardness H of the Al matrix, indentations in at least 30 positions (3 groups with 10 indentations each) for each alloy were carried out. A detailed description of the experimental conditions applied for nano-indentation can be found in [16].

In-situ tensile tests were conducted at the beamline ID19 of the European Synchrotron Radiation Facility (ESRF) in Grenoble, France [25], using a tensile rig with elevated temperature capabilities provided by INSA Lyon, Lyon, France. Flat dog-bone shaped tensile samples with a total length of 40 mm and 1 mm² cross-section at the gauge length of 2 mm were produced by spark erosion. Tensile tests were conducted at 300 °C at a strain rate of 1 μm/s and with controlled heating of the specimen central section by an induction coil monitored by a thermocouple glued to the sample. The test temperature was reached with a transient of a few seconds and was kept constant for the duration of the tensile test, with a maximum deviation of ±1 °C. Table 2 shows the experimental parameters for the sXCT scans carried out during the in-situ tensile tests at 300 °C. The first tomography was acquired before deformation and several tomographic scans were then subsequently acquired after applying increasing deformation steps until fracture. The determination of load steps of interest and the methodology for calculation of the global strain in the investigated volumes is described in detail in [16].

Table 2. Experimental parameters for tomography during the in-situ tensile tests at the beamline ID19/ European Synchrotron Radiation Facility (ESRF).

Experiment	Detector	Energy (keV)	FOV (mm^2)	Sample to Detector Distance (mm)	Exposure Time (s/proj)	Proj.	Voxel Size (μm^3)	Total Scan Time (s)
In-situ tensile tests at 300 °C	PCO Dimax	19	1 × 2	150	0.01	1000	1.1^3	20

The 3D microstructure of the alloys in the initial condition was also studied by sXCT using a higher spatial resolution than for the in-situ tensile tests. For this purpose, cylindrical specimens with diameters of 0.6 mm were machined. An exposure time of 0.02 s/projection and 4999 projections per scan were acquired with a voxel size of 0.3^3 μm^3.

2.3. Image Analysis

2.3.1. Pre-Processing

Reconstruction of tomographic scans was carried out with a filtered backprojection algorithm. The volumes at different load steps of the tensile tests were registered using the rigid registration tool available in Avizo Fire 9.3 (Thermo Scientific, Waltham, MA, USA). Prior to segmentation, the reconstructed volumes were filtered using a 2D or 3D anisotropic diffusion filter available in Fiji [26] and Avizo Fire 9.3.

2.3.2. Image Segmentation

Segmentation of aluminides was carried out using three different global grey value thresholds (best threshold determined by eye ±2 grey values) to increase the representativity of the quantitative analysis. An automatic segmentation of Si particles over a large volume was not possible by simple global grey value thresholding and, therefore, manual segmentation was carried out. 3D visualizations of regions of interest were produced using the software Avizo Fire.

2.3.3. Characterization of the 3D Microstructure and Damage

The quantification of the 3D microstructure of the alloys in the initial condition was determined in volumes of ~420 × 400 × 1180 μm^3. The interconnectivity of a phase, understood here as the global interconnectivity of that phase, was quantified as the volume of the largest particle divided by the total volume of all particles of the same phase in the studied volume, as defined in previous works [1,27]. Local connectivity, that is, connecting branches within a network, was quantified using the topological parameter Euler number χ [17,22–24]. A detailed explanation of the calculation of these parameters can be found in [16].

The morphology of primary Si was quantified aiming at identifying clusters formed by connected primary Si particles. For this, the aspect ratio and the sphericity, ψ, of primary Si particles were calculated as follows:

$$\text{aspect ratio} = \frac{\text{max.Feret} - \text{diameter}}{\text{min.Feret} - \text{diameter}}, \quad (1)$$

$$\Psi = \frac{6\pi^{1/2}v_{particle}}{s_{particle}^{\frac{3}{2}}}, \Psi \in [0,1], \tag{2}$$

where $V_{particle}$ is the particle volume in (μm^3) and $S_{particle}$ is the particle surface in (μm^2). A sphericity of 1 indicates spherical particles. A decrease in sphericity is accompanied by increasingly irregular particle shapes. On the other hand, an increase in aspect ratio indicates more elongated shapes.

For the characterization of damage during tensile deformation, a 3D Despeckle-filter (1 voxel), available in Avizo Fire, was applied to reduce artefacts after voids segmentation by global grey value thresholding. The smallest particle/void size considered for both the static and in-situ scans was 36 μm^3.

3. Results

3.1. Influence of Chemical Composition on the Microstructure

Initial Microstructure

Figure 2a,b shows the 2D microstructure of the investigated alloys. Both alloys show the presence of dendrites with eutectic/primary Si and intermetallic phases (aluminides) in the interdendritic region. Moreover, clusters of connected primary Si particles result in a rather heterogeneous distribution of this phase. There are also several intermetallic phases containing Si, Cu, Ni, Mg, Mn and Fe [20,28,29]. SEM micrographs of deep etched specimens reveal a large fraction of needle-like intermetallics in the 3 wt.% Ni alloy (see insert in Figure 2d). EDX mappings reveal that these phases correspond to Al-Cu-Ni-rich aluminides (Figure 2e), i.e., most probably δ-phase (Al$_3$CuNi) according to the literature (e.g., [29,30]).

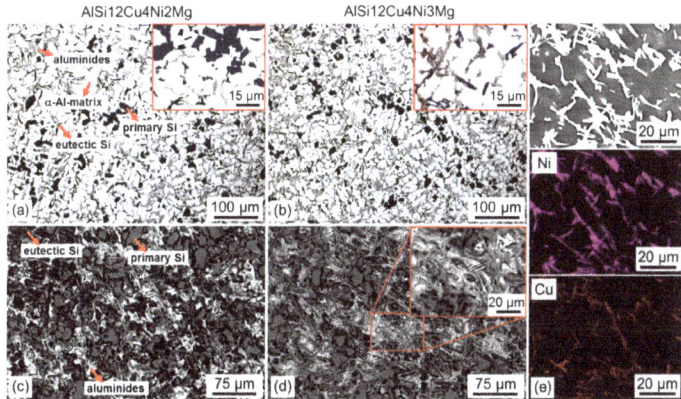

Figure 2. (a,b) light optical micrographs; (c,d) scanning electron microscopy (SEM) micrograph of the networks formed by Si and intermetallics after deep etching the α-Al matrix for AlSi12Cu4Ni2Mg (a,c) and AlSi12Cu4Ni3Mg (b,d); (e) SEM and energy-dispersive X-ray spectroscopy (EDX) maps of Ni and Cu for the AlSi12Cu4Ni3Mg alloy.

Table 3 presents average values for Brinell hardness of the alloys (HB) and nano-hardness of the α-Al matrix (H). Although the results are very similar considering the experimental scatter, the AlSi12Cu4Ni3Mg alloy may have a slightly higher Brinell hardness than AlSi12Cu4Ni2Mg.

Table 3. Brinell-hardness of the alloys and nano-hardness of the α-Al matrix.

Alloy	HB (1/10)	H (GPa)
AlSi12Cu4Ni2Mg	126 ± 2.7	1.8 ± 0.1
AlSi12Cu4Ni3Mg	129 ± 1.4	1.8 ± 0.1

Figure 3 displays portions of reconstructed sXCT slices of both alloys. Bright particles correspond to aluminides, while dark grey regions represent bulky primary Si particles and platelet-like eutectic Si.

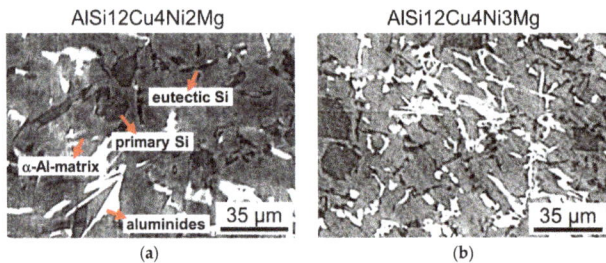

Figure 3. Portion of reconstructed synchrotron tomography slices: (**a**) AlSi12Cu4Ni2Mg and (**b**) AlSi12Cu4Ni3Mg.

Figure 4 shows visualizations of the 3D networks formed by intermetallics and eutectic + primary Si in a volume of 420 × 400 × 1180 µm^3. The left half of the visualization shows each segmented microstructural component separately: red = intermetallics, blue = primary Si particles, green = eutectic Si. On the other hand, the right halves of Figure 4a,b show all the rigid phases embedded in the transparent α-Al-matrix. Different colors have been assigned to each individual particle in this half of the volumes. While the aluminides (intermetallics) as well as primary + eutectic Si form globally highly interconnected 3D networks on their own (left half of each figure), all these phases together form a globally practically fully connected network in both alloys (right half of each figure) [1,16].

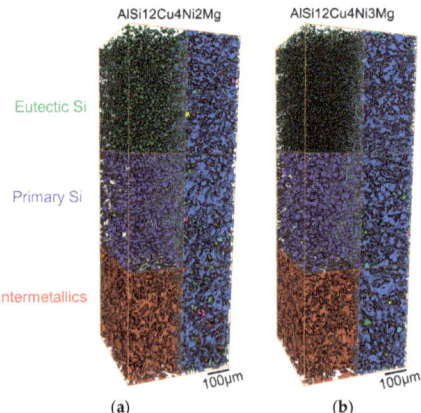

Figure 4. 3D visualization of the networks formed by the rigid phases in (**a**) AlSi12Cu4Ni2Mg and (**b**) AlSi12Cu4Ni3Mg. Left half of each figure: red = intermetallics, blue = primary Si particles, green = eutectic Si. Right half: network form considering all rigid phases together. Investigated volume: ~420 × 400 × 1180 µm^3.

The volume fraction, f, global interconnectivity and Euler number, χ, of Si and aluminides before deformation determined in volumes of 420 × 400 × 1180 µm³ are shown in Figure 5a,b individually and combined, that is, for the hybrid 3D network shown in Figure 4. Both alloys possess similar volume fractions of rigid phases (see Figure 5a) and—from a global point of view—highly interconnected networks with global interconnectivities of about 95% (see Figure 5b, green bars). A considerable difference between the 3D hybrid networks of the two alloys is given by their *local connectivity*, quantified here using the topological parameter Euler number χ (Figure 5b, red bars). The lowest Euler number ($-23{,}326 \pm -1115$) for the 3D network of AlSi12Cu4Ni3Mg indicates the highest local connectivity, that is, a larger number of connecting branches within the 3D network than for the alloy with 2 wt.% Ni, which shows about 41% less local connectivity ($\chi = -7277 \pm -436$).

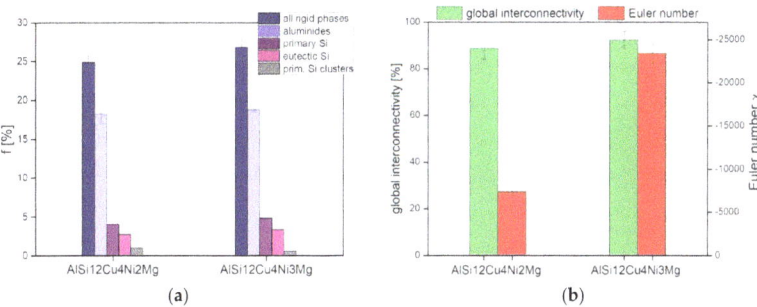

Figure 5. (**a**) Volume Fraction (f) of rigid phases, (**b**) global interconnectivity and Euler number χ, of the 3D network considering all rigid phases.

Large primary Si particles have frequently been reported to be favorable damage initiation sites during tensile deformation (e.g., [31–34]), however, the crucial effect of primary Si clusters on damage of near eutectic Al-Si alloys has received much less attention [35,36], thus, a detailed investigation of primary Si particles has been conducted. Figure 6 shows quantitatively the relationship between sphericity, aspect ratio and size of primary Si particles in the form of scatter diagrams. Color categories for several particle size ranges have been defined based on morphological changes observed qualitatively: (i) individual primary Si particles can be found for $V_{primary\ Si} \leq 3500$ µm³; (ii) small clusters of primary Si particles are present within 3500 µm³ $\leq V_{primary\ Si} \leq 7000$ µm³; (iii) larger clusters with up to about 15 connected particles are within 7000 µm³ $\leq V_{primary\ Si} \leq 10{,}000$ µm³; and (iv) very large, irregular clusters formed by more than 15 connected particles are found for $V_{primary\ Si} \geq 10{,}000$ µm³. Representative 3D visualizations of primary Si clusters for the size range(iv) are shown at the bottom of Figure 6 for the 2 wt.% Ni alloy (three images at the left hand side) and the 3 wt.% Ni alloy (3 images at the right hand side).

The sphericity evolves from rather spherical shapes at smaller particle sizes towards very irregular shapes with increasing size for both alloys. Clusters > 34,500 µm³ are heterogeneously distributed in the microstructure of both investigated alloys. The volume fraction of primary Si in the investigated volumes is 3.9 vol.% and 4.7 vol.% for the 2 wt.% Ni and the 3 wt.% Ni alloy, respectively. The larger volume fraction of primary Si can be attributed to the slightly higher Si content in the 3 wt.% Ni alloy (see Table 1). However, the volume fraction of the largest primary Si clusters in the investigated volumes ($V_{primary\ Si} > 34{,}500$ µm³) amounts to ~0.9 vol.% and ~0.5 vol.% for the 2 wt.% and 3 wt.% Ni alloy, respectively.

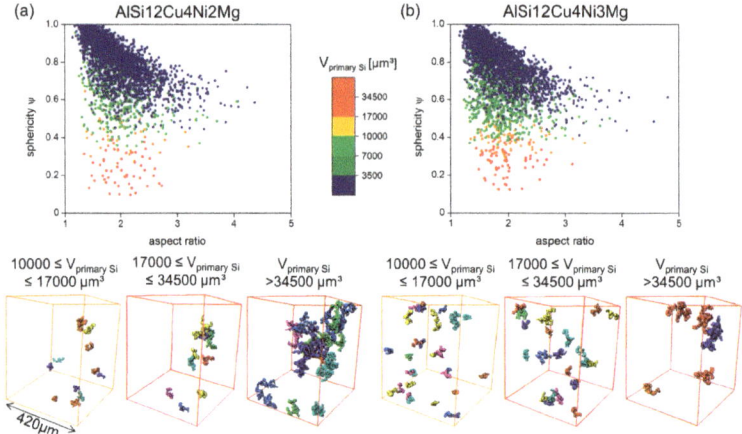

Figure 6. Sphericity-Aspect ratio-Volume diagrams for primary Si: (**a**) AlSi12Cu4Ni2Mg and (**b**) AlSi12Cu4Ni3Mg. The bottom of the figure shows representative 3D visualization of primary Si clusters for 10,000 µm^3 ≤ V$_{primary\ Si}$ ≤ 17,000 µm^3 (yellow), 17,000 µm^3 ≤ V$_{primary\ Si}$ ≤ 34,500 µm^3 (red) and V$_{primary\ Si}$ > 34,500 µm^3 (brown).

3.2. Tensile Tests at 300 °C

Figure 7 shows the stress-strain curves acquired during the in-situ tensile tests at 300 °C. The star-like symbols indicate the conditions at which sXCT was conducted. Both alloys show similar stress-strain evolution up to about 110 MPa. The 2 wt.% Ni alloy shows the largest elongation at failure (ε = 0.12 ± 0.006) and also the highest ultimate tensile strength σ_{UTS} = 141 MPa occurring at ε = 0.058 ± 0.002 compared to the 3 wt.% Ni alloy, for which σ_{UTS} = 132 MPa is reached already at ε = 0.0215 ± 0.005 and the elongation at fracture is ε = 0.086 ± 0.01.

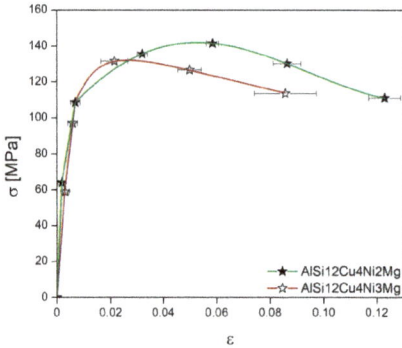

Figure 7. Stress-strain curves obtained during the in-situ tensile tests at 300 °C. The star-like symbols indicate the conditions at which synchrotron tomography was conducted.

Damage Formation and Accumulation during Tensile Tests at 300 °C

Figure 8a,b shows the damage (blue) at several deformation steps during tensile deformation at 300 °C as wells as in the post-mortem condition for the AlSi12Cu4Ni2Mg and AlSi12Cu4Ni3Mg alloys, respectively. The regions shown in yellow in the post-mortem conditions correspond to the fracture surface (the upper half of the fractured sample of the AlSi12Cu4Ni3Mg alloy could not be scanned and it is therefore missing in Figure 8b). It can also be seen qualitatively that AlSi12Cu4Ni2Mg shows a larger number and volume fraction of voids before final failure compared to AlSi12Cu4Ni3Mg, indicating that the alloy with less Ni is able to accommodate more damage before fracture. Damage forms and accumulates in a rather localized manner for both alloys. Moreover, AlSi12Cu4Ni2Mg shows the presence of randomly distributed processing porosity (see green particles in unloaded condition of this alloy in Figure 8a), which arises as a result of metal shrinkage during solidification. Damage was not observed to proceed from these shrinkage pores.

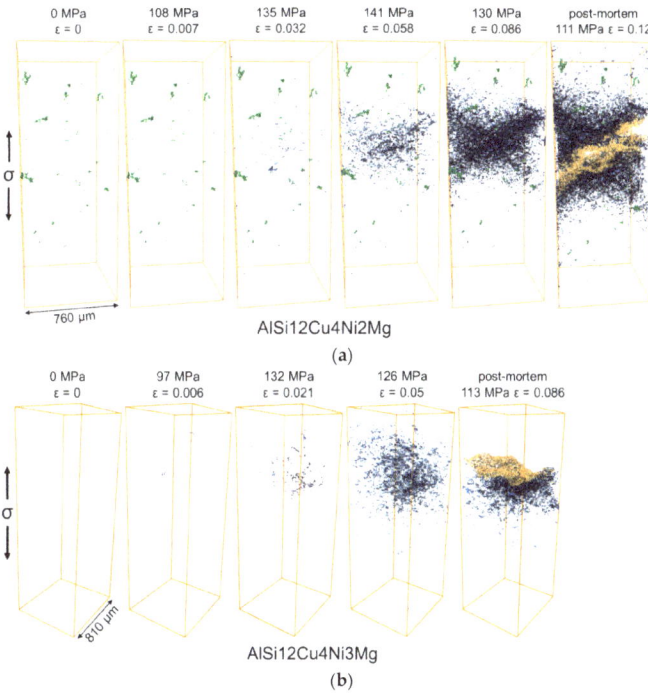

Figure 8. 3D visualization of processing porosity (green), voids (blue) near the fracture surface (beige) at several load steps during the in-situ tensile tests at 300 °C: (**a**) AlSi12Cu4Ni2Mg; (**b**) AlSi12Cu4Ni3Mg. The load direction is vertical.

The accumulation of damage is shown quantitatively in Figure 9 considering the number density (n_{voids}) and volume fraction (f_{voids}) of voids as a function of strain. Here, no distinction is made between micro-cracks and rounder voids. AlSi12Cu4Ni2Mg displays less damage than AlSi12Cu4Ni3Mg at the same applied strain and accumulates a larger fraction of damage prior to final failure.

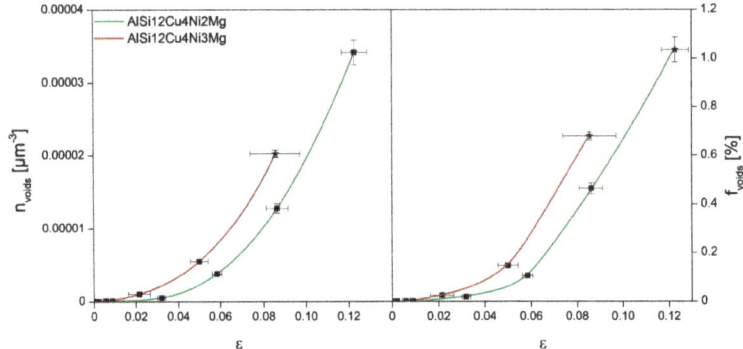

Figure 9. Number density (n_{voids}) and volume fraction (f_{voids}) of voids at several deformation steps.

Figure 10 shows representative portions of tomographic slices where the dominant damage mechanisms leading to fracture can be observed. Although the same damage mechanisms were observed for both alloys, damage was detected at earlier stages for the alloy with 3 wt.% Ni, that is, at σ = 97 MPa and ε = 0.006 ± 0.002 for AlSi12Cu4Ni3Mg, compared to σ = 135 MPa and ε = 0.032 ± 0.002 for AlSi12Cu4Ni2Mg, revealing the capability of the latter alloy to accommodate a larger amount of plastic deformation before damage initiation. Damage begins in the form of micro-cracks through primary Si particles agglomerated in clusters (red arrows in Figure 10). Round voids at the interfaces between α-Al-Matrix and the rigid phases were observed as well simultaneously for the AlSi12Cu4MgNi2 alloy, while this mechanism is detected at later stages of deformation for the alloy with 3 wt.% Ni (blue arrows in Figure 10). While the formation of cracks at primary Si particles forming cluster was also observed in our previous investigations during in-situ tensile tests of similar alloys at room temperature [16], the decohesion between matrix and rigid phases takes place only at elevated temperature. Lebyodkin et al. [34] reported that this damage mechanism is more likely to occur at elevated temperatures owing to relaxation of stresses at the interface between Al and Si. The cracks through Si particles are oriented preferentially perpendicular to the load direction.

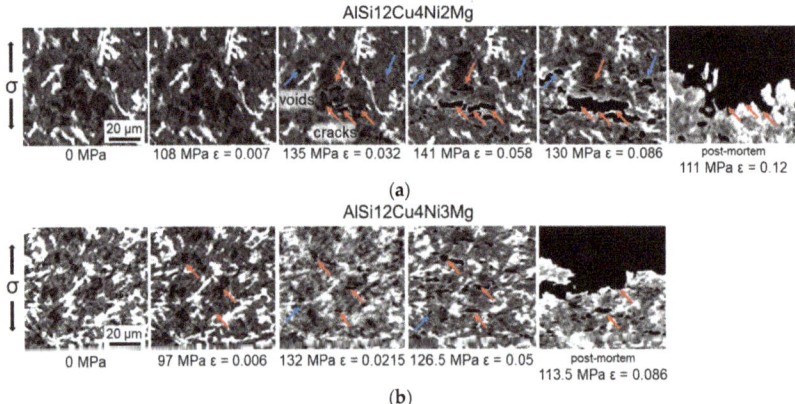

Figure 10. Damage mechanisms observed during the in-situ tensile tests: (a) AlSi12Cu4Ni2Mg; (b) AlSi12Cu4Ni3Mg. The load direction is vertical.

Figure 11 shows 3D visualizations of the damage mechanisms during deformation for volumes containing the 2D slices shown in Figure 10. Only the primary Si particles (blue) and an aluminide particle connected to the Si cluster are shown in this figure. Eutectic Si and further aluminides which are also present within this volume were made transparent to improve clarity. The micro-cracks through primary Si particles observed in the 2D slices result in a fragmentation of the originally fully connected particle cluster (blue) causing a localized damage accumulation in the early stages of failure. Furthermore, these 3D visualizations reveal that isolated Si particles can break if they are located close to clusters. It is also clear from these volumes that voids can also be formed in the α-Al-matrix, which is another distinction with respect to the damage mechanisms observed at room temperature (RT) [16] (some of the voids formed in the matrix while a fraction of them is located in eutectic Si and aluminides—or at their interface—transparent in this figure). Final failure occurs by coalescence of micro-cracks and voids, regardless of the alloy. Contrary to RT observations [16], where the main crack propagates exclusively along connecting micro-cracks formed at Si and aluminides, at 300 °C, propagation through voids in the α-Al-matrix is also observed (see the green crack, part of the main crack, shown in Figure 11). Even though damage mechanisms are the same for both alloys, the damage onset strain differs clearly between the alloys, as it can be seen in Figures 10 and 11.

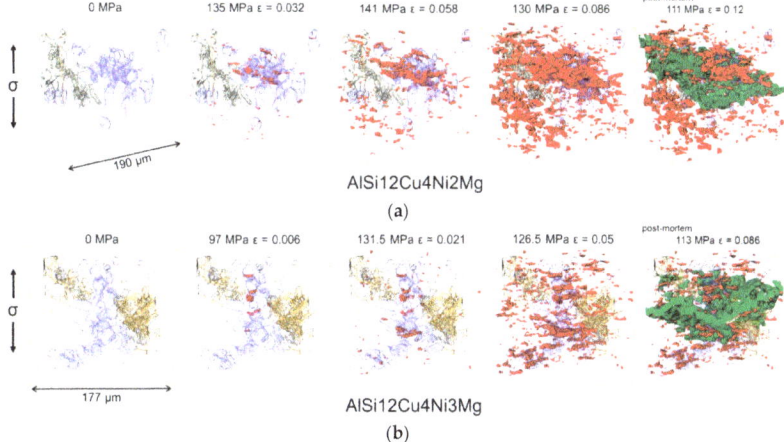

Figure 11. 3D visualization of damage accumulation at several load steps and in the post-mortem condition (main crack = green): (**a**) AlSi12Cu4Ni2Mg alloy and (**b**) AlSi12Cu4Ni3Mg alloy. Damage is shown in red, while primary Si particles are blue and aluminides beige. The matrix is transparent.

The local connectivity of the 3D networks (i.e., Euler number χ) changes as damage accumulates since this results in their fragmentation. Figure 12 shows the evolution of the Euler number of the 3D networks at several load steps calculated in the volumes shown in Figure 11 considering the full network of rigid phases contained in these volumes.

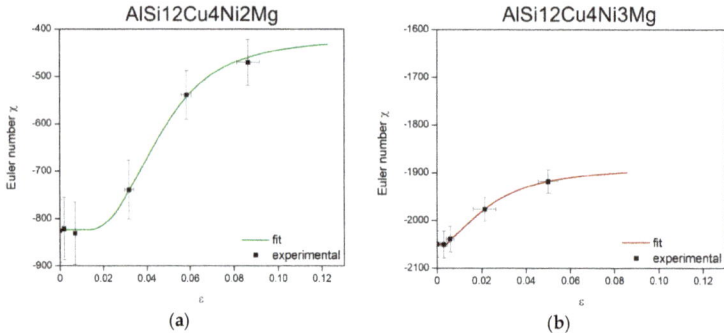

Figure 12. Evolution of Euler number of the 3D rigid network during tensile deformation for the volumes shown in Figure 11: (**a**) AlSi12Cu4Ni2Mg and (**b**) AlSi12Cu4Ni3Mg.

The Euler number remains constant until damage initiates. Then, an increase of χ with progressing deformation takes place owing to the loss of connecting branches by the formation of micro-cracks that progressively disintegrate the network. This increase of χ is more significant for the 2 wt.% Ni alloy (from initially −826 to −470 at the step before failure) in comparison to the 3 wt.% Ni alloy (from initially −2050 to −1886 at the step before failure). The experimentally determined evolution of the Euler number of the network as a function of strain, $\chi_n(\varepsilon)$, was fitted with an expression of the form:

$$\chi_n(\varepsilon) = A_2 + \frac{A_1 - A_2}{\left(1 + \left(\frac{\varepsilon}{x_0}\right)^p\right)}, \tag{3}$$

where A_1 is the initial value, A_2 is the final value, p is the power and x_0 is the center of the curve. The fitting parameters for each alloy are given in Table 4.

Table 4. Fit-parameters for Euler number curves in Figure 12.

Parameter	AlSi12Cu4Ni2Mg	AlSi12Cu4Ni3Mg
A_1	−822.5	−2051.4
A_2	−420.2	−1886.5
p	3.6	1.9
x_0	0.047	0.023

4. Discussion

4.1. Influence of Chemical Composition on the Microstructure and Damage Evolution

The sXCT results show that primary Si clusters play a major role in damage formation during tensile deformation at 300 °C, as it has also been observed at room temperature during a previous in-situ study of the tensile behavior of a similar alloy [16]. Both alloys studied in this work display large clusters of primary Si ($V_{\text{primary Si}} \geq 34{,}500$ µm^3, see Figure 6). It is known that Cu and Mg enhance the formation of dendrites by extending the solidification range of the alloys, which, in turn, allows the formation of primary Si chain-like clusters in the interdendritic space [6,29,37].

The damage mechanisms and their sequence identified during the in-situ tensile tests at 300 °C are very similar for both alloys:

1. Formation of micro-cracks through primary Si particles agglomerated in clusters, voids at matrix/rigid phase interfaces as well as voids in the matrix can be seen in the early stages of damage. Fracture of isolated primary Si particles and eutectic Si can also occur if they are

located in the vicinity of primary Si clusters. While the formation of voids in the matrix and at the interface between the matrix and the rigid phases was observed simultaneously with micro-cracking of primary Si in the AlSi12Cu4Ni2Mg alloy, these two mechanisms seem to occur at a later deformation stage for AlSi12Cu4Ni3Mg. Further sXCT scans at strains between 0.007 and 0.0032 are necessary for the AlSi12Cu4Ni2Mg alloy to fully clarify this difference.

2. Coalescence of voids leading to final failure with the main crack propagating along damaged rigid particles as well as through the matrix.

This is in agreement with previous reports in which an increase in the fraction of decohesion between Si/matrix interfaces and matrix damage in shape of voids was also observed after deformation of Al-Si alloys at elevated temperature [33,34,38–40]. Decohesion is known to result from accumulation of plastic deformation of the matrix in the vicinity of rigid phases, while fracture of rigid particles occurs owing to incompatible stresses between rigid phase and matrix [33]. Elevated temperatures facilitate plastic deformation of the matrix, and thus aid in alleviating the mismatch stresses, promoting decohesion instead of fracture [34]. Furthermore, a transition from brittle to ductile fracture modes with increasing test-temperature owing to crack propagation through the matrix has frequently been detected (e.g., [38,39]).

On the other hand, damage forms at earlier deformation steps for the 3% Ni alloy, which can be observed in Figures 10 and 11. The higher Ni content, results in the formation of plate-like δ-phase (see Figure 2b,d), resulting in the lowest Euler number ($-23,326 \pm -1115$) and thus, the highest local connectivity, for the 3D network of AlSi12Cu4Ni3Mg (see Figure 5b). The earlier damage initiation can be attributed to the increase of rigidity of the 3D network, restricting the plastic deformation of the matrix. The 3D network in AlSi12Cu4Ni2Mg has about 41% less local connectivity ($\chi = -7277 \pm -436$) compared to the 3 wt.% Ni alloy, that is, a considerably lower amount of connecting branches. This permits some plastic deformation of the matrix and helps accommodating more damage resulting in an increase of ductility with respect to AlSi12Cu4Ni3Mg, while the presence of more connecting branches in AlSi12Cu4Ni3Mg favors crack propagation through the rigid network and, consequently, final fracture occurs at lower strains with respect to the 2 wt.% Ni alloy.

4.2. Analytical Stress Partition Model to Gain Further Insights into the Mechanical Behavior of Al-Si Piston Alloys

It can be argued that a higher local connectivity increases the load carrying capability of the 3D networks, that is, more branches locally reinforcing the structure. Thus, a low Euler number (more negative) enhances the load transfer from the matrix to the rigid 3D network reducing the portion of load that must be borne by the matrix. In a first attempt to rationalize the effect of local connectivity on the load partition between matrix and the Si + aluminides 3D network, it is plausible to propose:

$$\frac{\sigma_m}{\sigma_n} = -\frac{C}{\chi}, \tag{4}$$

where σ_m is the load carried by the α-Al matrix, σ_n is the load carried by the rigid 3D network, C is a constant and χ is the Euler number. To determine C one can consider a condition at which deformation of the alloy is in the linear regime and damage has not yet initiated, e.g., the first deformation conditions for which tomography was carried out (see Figure 7). Here, the alloys can be assumed as continuously reinforced composites owing to the high *global interconnectivity* of the 3D networks, hence the matrix and the 3D network experience the same deformation:

$$\varepsilon_0 = \varepsilon_{n0} = \varepsilon_{m0}, \tag{5}$$

where ε_{n0} is the strain in the network and ε_{m0} is the strain in the matrix at ε_0. Approximating with the Hook's law:

$$\varepsilon_0 = \frac{\sigma_{m0}}{E_m} = \frac{\sigma_{n0}}{E_n}, \tag{6}$$

E_m can be approximated to the Young's Modulus of AA6061 at 300 °C (0.66 wt.% Si, 0.9 wt.% Mg, 0.24 wt.% Cu) [41]:

$$E_m = 50 \text{ GPa}, \tag{7}$$

Deriving from that, the matrix stress at the first deformation condition measured by sXCT results in $\sigma_{m0} = \sim 52$ MPa, based on the acquired stress-strain curves (Figure 7) with a practically identical elastic regime for both alloys.

The stress in the network at ε_0 can then be calculated as:

$$\sigma_{n0} = \frac{E_n \times \sigma_{m0}}{E_m}, \tag{8}$$

E_n can be approximated as:

$$E_n = f_{Si} \times E_{Si} + f_{aluminides} \times \overline{E}_{aluminides}, \tag{9}$$

where f_{Si} is 0.066 and 0.081 while $f_{aluminides}$ is 0.18 and 0.19 for the 2 wt.% Ni and the 3 wt.% Ni alloy, respectively (see Figure 5a). Several intermetallic phases must be considered to determine the Young's Modulus of the aluminides network $\overline{E}_{aluminides}$, therefore it is approximated here as the average of the Young's moduli measured at 200 °C for each present intermetallic phase taken from [42] (since the difference in Young's Modulus between these phases does not differ much—(± 6 GPa)—we assume the same fraction of phases for the calculation of the average Young's Modules $\overline{E}_{aluminides}$):

$$\overline{E}_{aluminides} = 144 \text{ GPa}, \tag{10}$$

The Young's Modulus of Si (E_{Si}) in dependence of temperature can be given in form of a linear equation approximated based on experimental data acquired in [42] in a temperature range 200 °C $\geq$ T $\leq$ 350 °C,

$$E_{Si}(T) = 125.9 - 0.07333 \times T, \tag{11}$$

According to this equation E_{Si} at 300 °C was estimated as,

$$E_{Si}(300\ °C) = 104 \text{ GPa}, \tag{12}$$

Using these values, E_n = 33 GPa and 36 GPa for the networks in AlSi12Cu4Ni2Mg and AlSi12Cu4Ni3Mg, respectively. With this, σ_{n0} = 36 MPa and 39 MPa for the 2 wt.% Ni alloy and the 3 wt.% Ni alloy, respectively. As a result, we can calculate C for each alloy,

$$C_{2\%} = -1251 \text{ Pa}, \tag{13}$$

$$C_{3\%} = -3184 \text{ Pa}, \tag{14}$$

The stress partition between the matrix and the 3D network can also be approximated by a simple rule of mixtures since, as previously mentioned, the high interconnectivity of the network permits an analogy with a continuously reinforced material:

$$\sigma - f_n \times \sigma_n + f_m \times \sigma_m = f_n \times \sigma_n + (1 - f_n) \times \sigma_m, \tag{15}$$

where f_m is the matrix volume fraction and f_n the volume fraction of the 3D network (Si + aluminides).

As mentioned in the results section, the local connectivity of the 3D networks, i.e., the Euler number, continuously changes during deformation owing to the formation and accumulation of damage. The evolution of Euler number as a function of applied strain, $\chi_n(\varepsilon)$, (see Figure 12) is used to implement the effect of local connectivity changes into the stress partition model. Combining Equations (3), (4) and (15) we can calculate the stress borne by the network as damage advances as:

$$\sigma_n = \sigma \times \frac{1}{\left(f_n + \frac{C}{\chi_n(\varepsilon)} \times (f_n - 1)\right)},\qquad(16)$$

The global stress σ is taken from the experimentally acquired stress-strain curves in Figure 7. The results of this approximation are plotted in Figure 13a,b for both alloys together with the evolution of χ of the 3D rigid networks and stress-strain curves from the in-situ tensile tests. The subscripts 2 and 3 are used for the 2 wt.% Ni and the 3 wt.% Ni alloy, respectively.

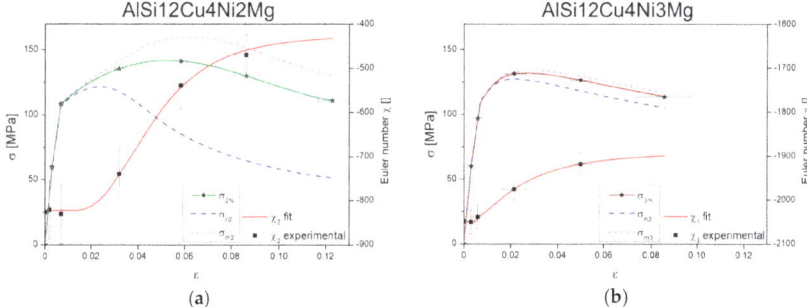

Figure 13. Evolution of stresses in the network and matrix calculated taking into account the disintegration of the 3D rigid networks by damage accumulation: (**a**) AlSi12Cu4Ni2Mg and (**b**) AlSi12Cu4Ni3Mg.

In the linear regime the network and matrix show the same evolution as the global stress-strain curves for both alloys. As damage begins, reflected by the increase of χ, the evolution of load borne by the matrices and the networks start to diverge. A more significant decrease of σ_n can be observed for the 2 wt.% Ni alloy after reaching a maximum at ~120 MPa, while, complementarily, σ_{m2} further increases by strain hardening up to ~160 MPa. On the other hand, σ_{n3} and σ_{m3} follow a very similar trend throughout the tensile test since damage is less pronounced and, thus, the local connectivity of the 3D network is less affected than for the alloy with 2 wt.% Ni. Moreover, the load carried by the 3D network of AlSi12Cu4Ni3Mg is always higher than for the alloy with less Ni in the non-linear regime, revealing the higher reinforcing capability of the hybrid network owing to a higher local connectivity. However, the high rigidity of the 3D network in the 3 wt.% Ni alloy hinders plastic deformation of the α-Al matrix and the possibility to further increase the strength of the alloy by strain hardening as deformation proceeds. σ_{n3} remains high and cracks in the network tend to coalesce and lead to earlier failure of the alloy with respect to the alloy with less Ni. Contrarily, the lower connectivity of the 3D network in AlSi12Cu4Ni2Mg has two consequences: first, its lower rigidity permits some plastic deformation and strain hardening of the α-Al matrix resulting in the higher strength of this alloy with respect to the alloy with 3 wt.% Ni and, second, more damage can be accommodated (in the matrix and at the interface between the matrix and the 3D network) leading to higher elongation at fracture than for AlSi12Cu4Ni3Mg.

This simple model, that takes into account the evolution of local connectivity of the 3D rigid networks as damage accumulates in cast Al-Si piston alloys, sheds a light on the effect of damage on strength of these materials and it is, to the best knowledge of the authors, the first attempt to rationalize the effect of global interconnectivity and local connectivity analytically using actual data obtained in-situ, during deformation at high temperature.

5. Conclusions

The influence of the 3D microstructure on the damage evolution of AlSi12Cu4Ni2Mg and AlSi12Cu4Ni3Mg cast piston alloys was studied during tensile deformation at 300 °C. The following conclusions are drawn:

- The additional formation of platelet-like Al-,Cu-,Ni-rich δ-phase owing to the increase of Ni content by 1 wt.% results in a larger amount of connecting branches and thus, a significant increase of *local connectivity* (quantified by the Euler number χ) of the rigid 3D network at practically constant *global interconnectivity* for the 3 wt.% Ni alloy.
- The in-situ tensile tests at 300 °C revealed ~10% higher strength and ~30% higher elongation at fracture of the 2 wt.% Ni alloy as compared to the 3 wt.% Ni alloy. Damage mechanisms during tensile deformation are the same for both alloys:

    ○ Damage mainly initiates as micro-cracking through primary Si particles agglomerated in clusters, voids at matrix/rigid phase interfaces, as well as voids in the matrix. Moreover, isolated primary Si particles and eutectic Si can break if they are located close to clusters. Failure occurs by coalescence of voids with the main crack propagating along fractured rigid particles as well as through the α-Al-matrix.
    ○ The lower local connectivity of the 3D network in AlSi12Cu4Ni2Mg permits local plastification of the matrix and helps accommodating more damage resulting in an increase of ductility with respect to AlSi12Cu4Ni3Mg. On the other hand, the 3 wt.% Ni alloy reveals damage onset at earlier deformation steps and less damage accumulation until failure compared to the 2% Ni alloy. The presence of more connecting branches in AlSi12Cu4Ni3Mg favors crack propagation through the rigid network and, consequently, final fracture occurs at lower strains with respect to the 2 wt.% Ni alloy.

- The evolution of local connectivity of the rigid 3D networks with damage accumulation based on experimental observations was implemented in a simple analytical stress partition model. The lower local connectivity at practically constant global interconnectivity of the 3D network in AlSi12Cu4Ni2Mg indicates two consequences:

    ○ its lower rigidity allows local plastification and strain hardening of the α-Al matrix resulting in the higher strength of this alloy.
    ○ more damage can be accommodated leading to higher elongation at fracture than for AlSi12Cu4Ni3Mg.

Author Contributions: K.B. performed the experiments, analysed the data, contributed to interpretation and discussion and wrote the paper; G.R. conceived and designed the project, contributed to interpretation, discussion and writing of the paper; T.S. and H.G. contributed to design the project, produced and provided the material, contributed to discussion of results; E.M. and J.A. provided the in-situ test rig, provided support during the in-situ experiments; F.S. performed the in-situ experiments, contributed to discussion of results; E.B. support during the in-situ experiments and 3D data reconstruction.

Funding: This work is part of the "K-Project for Non-Destructive Testing and Tomography Plus" supported by the COMET-Program of the Austrian Research Promotion Agency (FFG) as well as the Provinces of Upper Austria (LOÖ) and Styria, Grant No. 843540.

Acknowledgments: The ESRF is acknowledged for the provision of synchrotron facilities at the beamline ID19.

Conflicts of Interest: The authors declare no conflict of interest.

References

1. Asghar, Z.; Requena, G.; Boller, E. 3D rigid multiphase networks providing high-temperature strength to cast AlSi10Cu5Ni1-2 piston alloys. *Acta Mater.* **2011**, *59*, 6420–6432. [CrossRef] [PubMed]
2. Requena, G.; Garcés, G.; Asghar, Z.; Marks, E.; Staron, P.; Cloetens, P. The Effect of Connectivity of Rigid Phases on Strength of Al-Si Alloys. *Adv. Eng. Mater.* **2011**, *13*, 674–684. [CrossRef]
3. Mohamed, A.M.A.; Samuel, F.H. A Review on the Heat Treatment of Al-Si-Cu/Mg Casting Alloys. *INTEC* **2012**. [CrossRef]
4. Zhu, P.Y.; Liu, Q.Y.; Hou, T.X. Spheroidization of eutectic Silicon in Aluminum-Silicon Alloys. *AFS Trans.* **1985**, *93*, 609–614.
5. Konečná, R.; Fintová, S.; Nicoletto, G.; Riva, E. High temperature fatigue strength and quantitative metallography of an eutectic Al-Si alloy for piston application. *Key Eng. Mater.* **2014**, *592–593*, 627–630. [CrossRef]
6. Shabestari, S.G.; Moemeni, H. Effect of copper and solidification conditions on the microstructure and mechanical properties of Al-Si-Mg alloys. *J. Mater. Process. Technol.* **2004**, *153–154*, 193–198. [CrossRef]
7. Efzan, M.N.E.; Kong, H.J.; Kok, C.K. Review: Effect of Alloying Element on Al-Si alloys. *Adv. Mater. Res.* **2014**, *845*, 355–359. [CrossRef]
8. Jeong, C.-Y. Effect of Alloying elements on high temperature mechanical properties for piston alloy. *Mater. Trans.* **2012**, *53*, 234–239. [CrossRef]
9. Zhang, W.D.; Yang, J.; Dang, J.Z.; Liu, Y.; Xu, H. Effects of Si, Cu and Mg on the high temperature mechanical properties of Al-Si-Cu-Mg alloy. *Adv. Mater. Res.* **2013**, *652–654*, 1030–1034. [CrossRef]
10. Silva, F.S. Fatigue on engine pistons—A compendium of case studies. *Eng. Fail. Anal.* **2006**, *13*, 480–492. [CrossRef]
11. Asghar, Z.; Requena, G.; Kubel, F. The role of Ni and Fe aluminides on the elevated temperature strength of an AlSi12 alloy. *Mater. Sci. Eng. A* **2010**, *527*, 5691–5698. [CrossRef]
12. Asghar, Z.; Requena, G.; Degischer, H.P.; Cloetens, P. Three-dimensional study of Ni aluminides in an AlSi12 alloy by means of light optical and synchrotron microtomography. *Acta Mater.* **2009**, *57*, 4125–4132. [CrossRef]
13. Requena, G.; Degischer, H.P. Three-Dimensional Architecture of Engineering Multiphase Metals. *Ann. Rev. Mater. Res.* **2012**, *42*, 145–161. [CrossRef]
14. Asghar, Z.; Requena, G.; Zahid, G.H. Effect of thermally stable Cu- and Mg-rich aluminides on the high temperature strength of an AlSi12CuMgNi alloy. *Mater. Charact.* **2014**, *88*, 80–85. [CrossRef]
15. Asghar, Z.; Requena, G.; Degischer, H.P.; Boller, E. Influence of Internal Architectures of Cast AlSi10Cu5Ni1-2 on High Temperature Strength. In Proceedings of the 12th International Conference on Aluminium Alloys, Yokohama, Japan, 5–9 September 2010; pp. 1285–1290.
16. Bugelnig, K.; Sket, F.; Germann, H.; Steffens, T.; Koos, R.; Wilde, F.; Boller, E.; Requena, G. Influence of 3D connectivity of rigid phases on damage evolution during tensile deformation of an AlSi12Cu4Ni2 piston alloy. *Mater. Sci. Eng. A* **2018**, *709*, 193–202. [CrossRef]
17. Kruglova, A.; Engstler, M.; Gaiselmann, G.; Stenzel, O.; Schmidt, V.; Roland, M.; Diebels, S.; Mücklich, F. 3D connectivity of eutectic Si as key property defining strength of Al-Si alloys. *Comput. Mater. Sci.* **2016**, *120*, 90–107. [CrossRef]
18. Cho, Y.-H.; Im, Y.-R.; Kwon, S.-W.; Lee, H.C. The Effect of Alloying Elements on the Microstructure and Mechanical Properties of Al-12Si Cast Alloys. *Mater. Sci. Forum* **2003**, *426–432*, 339–344. [CrossRef]
19. Stadler, F.; Antrekowitsch, H.; Fragner, W.; Kaufmann, H.; Uggowitzer, P.J. The effect of Ni on the high-temperature strength of Al-Si cast alloys. *Mater. Sci. Forum* **2011**, *690*, 274–277. [CrossRef]
20. Konečná, R.; Nicoletto, G.; Kunz, L.; Svoboda, M.; Bača, A. Fatigue strength degradation of AlSi12cuNiMg alloy due to high temperature exposure: A structural investigation. *Procedia Eng.* **2014**, *74*, 43–46. [CrossRef]
21. Li, Y.; Yang, Y.; Wu, Y.; Liu, X. Quantitative comparison of three Ni-containing phases to the elevated-temperature properties of Al–Si piston alloys. *Mater. Sci. Eng. A* **2010**, *527*, 7132–7137. [CrossRef]
22. Toriwaki, J.; Yonekura, T. Euler Number and Connectivity Indexes of a Three Dimensional Digital Picture. *Forma* **2002**, *17*, 183–209.
23. Aydogan, D.B.; Hyttinen, J. Characterization of microstructures using contour tree connectivity for fluid flow analysis. *R. Soc. Interface* **2014**, *11*. [CrossRef] [PubMed]

24. Silva, M.J.; Gibson, L.J. The effect of non-periodic microstructure and defects on the compressive strength of two-dimensional cellular solids. *Int. J. Mech. Sci.* **1997**, *39*, 549–563. [CrossRef]
25. ID19—Microtomography Beamline. Available online: http://www.esrf.eu/home/UsersAndScience/Experiments/StructMaterials/ID19.html (accessed on 5 July 2018).
26. Schindelin, J.; Arganda-Carreras, I.; Frise, E. Fiji: An open-source platform for biological-image analysis. *Nat. Methods* **2012**, *9*, 676–682. [CrossRef] [PubMed]
27. Requena, G.C.; Degischer, P.; Marks, E.D.; Boller, E. Microtomographic study of the evolution of microstructure during creep of an AlSiCuMgNi alloy reinforced with Al2O3 short fibres. *Mater. Sci. Eng. A* **2008**, *487*, 99–107. [CrossRef]
28. Zeren, M. Effect of copper and silicon content on mechanical properties in Al-Cu-Si-Mg alloys. *J. Mater. Process. Technol.* **2005**, *169*, 292–298. [CrossRef]
29. Belov, N.A.; Eskin, D.G.; Avxentieva, N.N. Constituent phase diagrams of the Al-Cu-Fe-Mg-Ni-Si system and their application to the analysis of aluminium piston alloys. *Acta Mater.* **2005**, *53*, 4709–4722. [CrossRef]
30. Farkoosh, A.; Javidani, M.; Hoseini, M.; Larouche, D.; Pekguleryuz, M. Phase formation in as-solidified and heat-treated Al-Si-Cu-Mg-Ni alloys, Thermodynamic assessment and experimental investigation for alloy design. *J. Alloy Compd.* **2013**, *551*, 596–606. [CrossRef]
31. Ceschini, L.; Boromei, I.; Morri, A.; Seifeddine, S.; Svensson, I.L. Microstructure, tensile and fatigue properties of the Al-10%Si-12%Cu alloy with different Fe and Mn content under controlled conditions. *J. Mater. Process. Technol.* **2009**, *209*, 5669–5679. [CrossRef]
32. Zamani, M.; Seifeddine, S.; Jarfors, A.E.W. High Temperature tensile deformation and failure mechanisms of an Al-Si-Cu-Mg cast alloy—The microstructural scale effect. *Mater. Des.* **2015**, *86*, 361–370. [CrossRef]
33. Su, J.F.; Nie, X.; Stoilov, V. Characterization of fracture and debonding of Si particles in AlSi Alloys. *Mater. Sci. Eng. A* **2010**, *527*, 7168–7175. [CrossRef]
34. Lebyodkin, M.; Deschamps, A.; Bréchet, Y. Influence of second phase morphology and topology on mechanical and fracture properties of Al-Si alloys. *Mater. Sci. Eng. A* **1997**, *234–236*, 481–484. [CrossRef]
35. Gall, K.; Horstemeyer, M.; McDowell, D.L.; Fan, J. Finite element analysis of the stress distributions near damaged Si particle clusters in cast Al-Si alloys. *Mech. Mater.* **2000**, *32*, 277–301. [CrossRef]
36. Morgenstern, R.; Kenningley, S. Transient microstructural thermomechanical fatigue and deformation characteristics under superimposed mechanical and thermal loading, in AlSi based automotive diesel pistons. In *Light Metals 2013*; The Minerals, Metals & Materials Series; Sadler, B.A., Ed.; Springer: Cham, Switzerland, 2016; pp. 397–403. ISBN 978-3-319-65136-1.
37. Mondolfo, L.F. *Aluminium Alloys: Structure and Properties*; Butterworth & Co. Ltd.: London, UK, 1976; ISBN 0408709324.
38. Joyce, M.R.; Styles, C.M.; Reed, P.A.S. Elevated temperature short crack fatigue behaviour in near eutectic Al-Si alloys. *Int. J. Fatigue* **2003**, *25*, 863–869. [CrossRef]
39. Zhang, G.; Zhang, J.; Li, B.; Cai, W. Characterization of tensile fracture in heavily alloyed Al-Si piston alloy, Progress in Natural Science. *Mater. Int.* **2011**, *21*, 380–385.
40. Wang, M.; Pang, J.; Qiu, Y.; Liu, H.; Li, S.; Zhang, Z. Tensile Strength Evolution and Damage Mechanisms of Al–Si Piston Alloy at Different Temperatures. *Adv. Eng. Mater.* **2018**, *20*, 1700610. [CrossRef]
41. Summers, P.T.; Chen, Y.; Rippe, C.M.; Allen, B.; Mouritz, A.P.; Case, S.W.; Lattimer, B.Y. Overview of aluminum alloy mechanical properties during and after fires. *Fire Sci. Rev.* **2015**, *4*, 3. [CrossRef]
42. Chen, C.L.; Richter, A.; Thomson, R.C. Investigation of mechanical properties of intermetallic phases in multi-component Al-Si using hot-stage nanointendation. *Intermetallics* **2010**, *18*, 499–508. [CrossRef]

© 2018 by the authors. Licensee MDPI, Basel, Switzerland. This article is an open access article distributed under the terms and conditions of the Creative Commons Attribution (CC BY) license (http://creativecommons.org/licenses/by/4.0/).

Article

Strengthening of Aluminum Wires Treated with A206/Alumina Nanocomposites

David Florián-Algarín [1], Raúl Marrero [2], Xiaochun Li [3], Hongseok Choi [4] and Oscar Marcelo Suárez [5,*]

1. Department of Civil Engineering, University of Puerto Rico-Mayagüez, Mayagüez, PR 00681, USA david.florian@upr.edu
2. Department of Civil and Environmental Engineering, Northwestern University, Evanston, IL 60208, USA; raulmarrero2015@u.nortwestern.edu
3. Department of Mechanical and Aerospace Engineering, University of California-Los Angeles, Los Angeles, CA 90095-1597, USA; xcli@seas.ucla.edu
4. Department of Mechanical Engineering, Clemson University, Clemson, SC 29634, USA; hongc@clemson.edu
5. Department of Engineering Science and Materials, University of Puerto Rico-Mayagüez, Mayagüez, PR 00681, USA
* Correspondence: oscarmarcelo.suarez@upr.edu

Received: 9 February 2018; Accepted: 9 March 2018; Published: 10 March 2018

Abstract: This study sought to characterize aluminum nanocomposite wires that were fabricated through a cold-rolling process, having potential applications in TIG (tungsten inert gas) welding of aluminum. A206 (Al-4.5Cu-0.25Mg) master nanocomposites with 5 wt % γAl_2O_3 nanoparticles were first manufactured through a hybrid process combining semi-solid mixing and ultrasonic processing. A206/1 wt % γAl_2O_3 nanocomposites were fabricated by diluting the prepared master nanocomposites with a monolithic A206 alloy, which was then added to a pure aluminum melt. The fabricated Al–γAl_2O_3 nanocomposite billet was cold-rolled to produce an Al nanocomposite wire with a 1 mm diameter and a transverse area reduction of 96%. Containing different levels of nanocomposites, the fabricated samples were mechanically and electrically characterized. The results demonstrate a significantly higher strength of the aluminum wires with the nanocomposite addition. Further, the addition of alumina nanoparticles affected the wires' electrical conductivity compared with that of pure aluminum and aluminum–copper alloys. The overall properties of the new material demonstrate that these wires could be an appealing alternative for fillers intended for aluminum welding.

Keywords: aluminum nanocomposites; aluminum welding; TIG fillers; electrical conductivity; wire fabrication

---

1. Introduction

Welding, as a critical technique to join structural parts, requires a reliable filling material [1]. In particular, welding Al alloys with a proper filling material has become more important in setting up lightweight structures, such as those in aerospace applications [1–5]. Therefore, tuning the mechanical and thermal properties of the filler material is of vital importance for the welding quality. In effect, high strength prevents failure of the weld, eventually enhancing the performance of the final product. In terms of the weld thermal properties, the melting temperature of a welding material should be lower than that of the parts to be joined. One needs to recall that as the metal strength levels increase, the melting point is normally higher. This means that the balance between the strength and the melting point of the welding materials is of great importance.

Metal matrix composites bearing nanostructured components (nanoparticles, nanofibers, nanotubes, or graphene) embedded in a light metallic matrix have been considered for a number of applications due to their low density, high strength, and high elastic modulus, among other properties. For instance, ceramic nanoparticles, such as silicon nitride (Si_3N_4), silicon carbide (SiC), zirconia (ZrO_2), and alumina (Al_2O_3), are reinforcements utilized in nanocomposites because of their high temperature strength, high wear resistance, chemical stability, and large elastic modulus [6,7]. Among those ceramic nanoparticles, γAl_2O_3 is an effective reinforcement for aluminum due to its high hardness, mechanical strength, and good thermal shock resistance, allowing for metal matrix composites with high mechanical strength [8,9]. In addition, most engineering alumina parts come from γAl_2O_3 sintering. The ensuing phase development is then: $\gamma Al_2O_3 \rightarrow \delta Al_2O_3 \rightarrow \theta Al_2O_3 \rightarrow \alpha Al_2O_3$ [10,11]. Therefore, the cost of the δ, θ, and α phases is higher compared to the γ polymorph. Synthesized as nanoparticles, γAl_2O_3 can then be added as nanodispersoids to formulate an aluminum matrix nanocomposite, i.e., a nanoreinforced Al alloy.

It is apparent that effective incorporation of the ceramic particles into the aluminum (or aluminum alloy) matrix is paramount to obtaining a homogeneous and sound composite, free of pores or a weak ceramic/matrix interface (for load transfer purposes). The methods described above are effective but most of the time it is costly to attain a strong reinforcement/matrix interface. In this respect, reactive mixing could be an alternative to form a strong ceramic/metal interface, although at a price [12]. Moreover, as the size of the reinforcements becomes smaller (nanocomposites), the potential for agglomeration becomes higher. Thus, we propose the use of an aluminum alloy with well-embedded nanoparticles as a master material to inoculate an aluminum melt so as to formulate an Al matrix composite with superior strength. Such is the target of the present research, which evaluates the fabrication of a nanocomposite containing alumina nanoparticles (incorporated as part of an Al–4.5Cu–0.25Mg–1% γAl_2O_3 composite). The effect of its addition on the physical properties of aluminum wires, namely electrical resistivity, density, and melting point, are also studied, along with the stiffness and strength of the reinforced material. Our ultimate purpose is to produce a filler material for potential use in aluminum welding via a TIG (tungsten inert gas) method; this new material could be suitable for aerospace applications, specifically for structural joints.

2. Experimental Procedure

To produce the wires, three stages were necessary. The first stage was the manufacture of a master nanocomposite bearing an A206 alloy matrix [13] and containing γAl_2O_3 nanoparticles. This master composite was used then to inoculate an aluminum melt to attain a metal with a smaller amount of nanoparticles. The last stage was the wire drawing and subsequent characterization of its microstructure, electrical resistivity, thermal behavior, and mechanical strength.

2.1. Fabrication of a Master Al Matrix Nanocomposite

Table 1 shows the chemical composition of the A206 alloy used as the master composite matrix. An initial melt with 27 kg of A206 and 1 wt % of γAl_2O_3 was prepared using semi-solid mixing and ultrasonic processing. To enhance the semi-solid mixing of the nanoparticles, we used an axial impeller with a 25.4 mm diameter placed at a third of the total height of the crucible containing the A206 slurry. As the impeller rotated at 500 rpm, it formed a vortex into which 1.6 g of γAl_2O_3 nanoparticles (with an average 50 nm size) was added. Thereupon, we raised the impeller angular velocity to 1200 rpm and set the mixing time at 40 s. This feeding and mixing step was repeated until 5 wt % of nanoparticles had been added to the melt.

Table 1. A206 nominal chemical composition [13].

Alloy	wt % Cu	wt % Mn	wt % Mg	wt % Ti
A206.0	4.5	0.3	0.25	0.22

After the semi-solid mixing at 630 °C, we raised the temperature (700 °C) to enhance the distribution and dispersion of the nanoparticles via ultrasonic processing. To this purpose, the tip of a niobium (C-103) ultrasonic probe with a diameter of 12.7 mm and a length of 92 mm was inserted about 6.35 mm into the melt. A 20 kHz ultrasonic vibration bearing an amplitude (peak-to-peak) of 60 µm was generated by a transducer (Sonicator 3000, Misonix Inc., Farmingdale, NY, USA) and applied to the melt for 15 min. Then, the melt was heated to 740 °C for pouring into low-carbon steel rectangular molds preheated at 350 °C to obtain master nanocomposite ingots weighing 500 g. A similar ultrasonic processing was used in prior research where more details are provided [14].

The prepared A206/5 wt % γAl_2O_3 master nanocomposites were diluted with as-received A206 alloy to fabricate A206/1 wt % γAl_2O_3 nanocomposite castings. The scale-up ultrasonic processing system and a grade 5 titanium alloy ultrasonic probe were used for 30 min to further disperse and distribute the nanoparticles. After that, the melt was cast into permanent and sand molds. Standard A206 aluminum alloy is known for its hot tearing trend, which can be counteracted by adding γAl_2O_3 nanoparticles as proven in prior research [15,16]. Because the parent ingots needed to be tested for integrity in order to produce convenient inoculation shapes (i.e., wires), after pouring the melt, the hot tearing susceptibility of the nanocomposites was assessed by a constrained rod casting (CRC) with a steel mold, similar to those used in prior work [17].

2.2. Wire Fabrication and Characterization

This research segment involved the production of a nanocomposite bearing a matrix made of aluminum with 4.5 wt % Cu and 0.25 wt % Mg. We used 1 wt % γAl_2O_3 nanoparticles as reinforcement. To this purpose, pure aluminum (99.5%) was melted at 760 °C and the A206/1 wt % γAl_2O_3 nanocomposite was added to the mechanically stirred melt.

We poured the treated melt into a cylindrical mold to produce 6 mm diameter ingots. These underwent full annealing at 400 °C for 5 h to allow cold-rolling to obtain 2.6 mm diameter wires with a cross-sectional area reduction of 81%. Full annealing for 5 h at 400 °C permitted further cold rolling to reduce the wire diameter to 1.4 mm. Finally, another full annealing at 400 °C for 5 h allowed for a 1 mm wire diameter, which was used in prior research [18,19]. The standard tensile tests at room temperature took place in a low force universal testing machine, Instron 5944, following the ASTMB557-06 standard [20]. The fractured wire surfaces were observed in a JEOL SEM-6390 scanning electron microscope (SEM, Tokyo, Japan). We cut, ground, and polished the wires to observe their microstructure (wires at different stages of the manufacturing process) in a Nikon Epiphot 200 optical microscope (Tokyo, Japan).

A four-point probe technique allowed the electrical resistivity of the wires to be measured at different temperatures [21]. The apparatus measured the voltage drop between two probing points as different levels of current were applied to the wire via two electrodes. Then, by measuring the sample geometry, we computed the bulk material conductivity. The wires' electrical conductivity measurements were carried out at temperatures ranging from 0 to 100 °C to simulate some operating conditions. Finally, we measured the electrical conductivity as a percent of IACS, as the International Annealed Copper Standard (IACS) indicates [22]. Tables 1 and 2 present the target chemical composition of the A206 alloy and the aluminum wires treated with the A206/1 wt % γAl_2O_3 nanocomposite. Four specimens of each composition were manufactured.

Table 2. Wires chemical composition.

Wires	wt % Al	wt % γAl_2O_3	wt % Cu	wt % Mn	wt % Mg	wt % Ti
Al-12.5 wt % (A206/1 wt % γAl_2O_3)	99.218	0.125	0.562	0.037	0.031	0.027
Al-25.0 wt % (A206/1 wt % γAl_2O_3)	98.433	0.250	1.125	0.075	0.062	0.055
Al-37.5 wt % (A206/1 wt % γAl_2O_3)	97.651	0.375	1.687	0.112	0.093	0.082
Al-50.0 wt % (A206/1 wt % γAl_2O_3)	96.865	0.500	2.250	0.150	0.125	0.110

Due to the low Mn, Mg, and Ti levels in the wires, we opted to disregard their effects on the properties of the wires to be studied. Conversely, we were interested in assessing how different amounts of γAl$_2$O$_3$ and Cu could influence those properties. To differentiate between the effects of γAl$_2$O$_3$ and Cu in mechanical and electrical responses, aluminum–copper wires were fabricated bearing 0.562 wt % Cu, 1.125 wt % Cu, 1.687 wt % Cu, and 2.250 wt % Cu, using the same procedure, and an Al-33 wt % Cu (eutectic composition) master alloy as the charge material.

3. Results

3.1. Master A206/γAl$_2$O$_3$ Composite Characterization

Figure 1 shows two optical micrographs of the fabricated master nanocomposite alloy, obtained at different magnifications [16]. The images demonstrate that the γAl$_2$O$_3$ nanoparticles were well incorporated into the matrix and nearby the θ (Al$_2$Cu)/aluminum eutectic regions. Although it is also observed that the γAl$_2$O$_3$ nanoparticle clusters ended up pushed into interdendritic regions, i.e., last regions to solidify, those small agglomerates would be further dispersed and distributed during the subsequent scale-up ultrasonic processing, as can be seen in Figure 2. It should be noted that 1 wt % of γAl$_2$O$_3$ nanoparticles, which would serve as reinforcements in the master nanocomposite alloy, were successfully incorporated into the matrix through the hybrid mixing method.

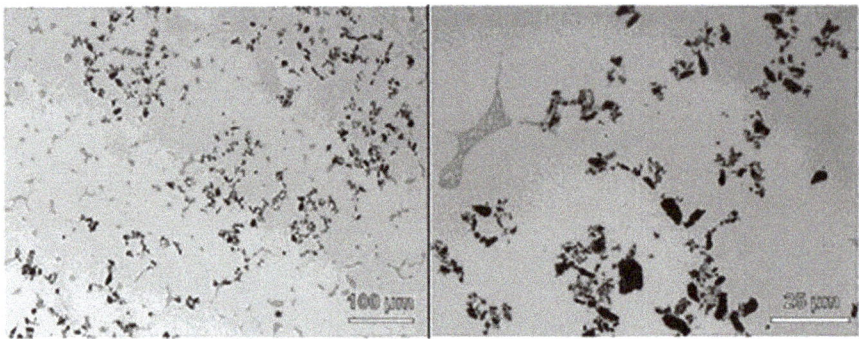

Figure 1. Optical micrographs of the A206/5 wt % γAl$_2$O$_3$ master nanocomposite alloy obtained at two magnifications (Reproduced with permission of the ASME from [16]).

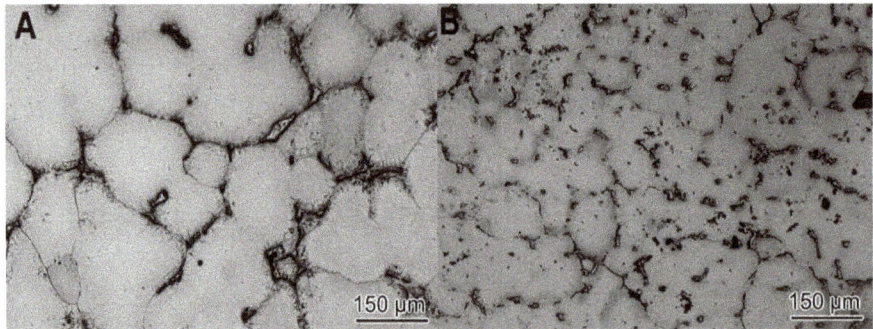

Figure 2. Optical micrographs of the alloy. (**A**) A206 aluminum alloy; (**B**) A206/1 wt % γAl$_2$O$_3$ master nanocomposite alloy.

This is an important starting point since the nanocomposite was intended to be used as the γAl_2O_3 nanoparticle carrier for the posterior inoculation of the aluminum melt. The small agglomeration was, therefore, a non-issue since the clusters were uniformly distributed throughout the parent material. The hot tearing susceptibility of the nanocomposites was compared with that of pure A206 alloy. As shown in Figure 3, the hot tearing resistance of the A206/1 wt % γAl_2O_3 nanocomposite was significantly better than that of the pure A206 alloy. The pure A206 casting evinced severe tears (marked with arrows) all around the part; conversely, the nanocomposite had only small tears (marked with circles) and partially around the part. This ensured the integrity of the ingots and warranted the soundness of the wires for plastic deformation in an ensuing stage of the processing.

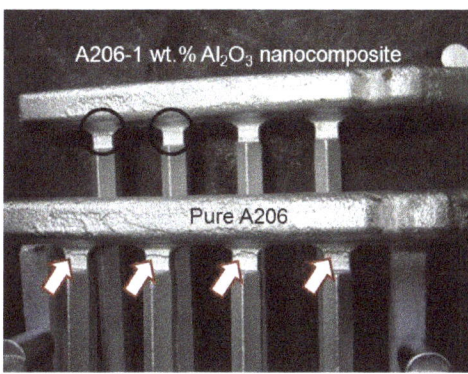

Figure 3. Comparison of pure A206 and A206/1 wt % γAl_2O_3 nanocomposite.

3.2. Wire Characterization

We deem it important to underscore that a rigorous characterization of the reinforced wires paves the way to their potential applications in the aforementioned industries. Therefore, a purely mechanical study would not suffice to evaluate the compliance of this new material with industrial specifications.

3.2.1. Tensile Test Results

Figures 4 and 5 present the ultimate tensile strength (UTS) results; the strain rate was 1 mm/min and the initial length of the wires was 250 mm, following the ASTMB557-06 standard. Clearly, the UTS increased as the amount of A206/1 wt % γAl_2O_3 (i.e., the amount of nanoparticles) and copper increased. Naturally, when nanoparticles are present, the increase in the UTS is attributed to Orowan strengthening mechanisms. The isotropic and uniform distribution of the nanoparticles present in the said matrix provided effective obstacles against dislocation slippage upon plastic deformation. Similar trends have been well documented when SiC, Al_2O_3, Y_2O_3, SiO_2, and carbon nanotube particles are embedded in pure magnesium and Mg alloys as matrix material [23]. In aluminum alloys, to increase the mechanical strength, one can use solid solution, cold working with or without heat treatment, as well as second phase precipitates and nanodispersions or nanoparticles, as in our case [24].

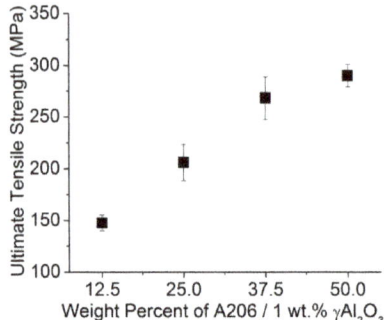

Figure 4. Average ultimate tensile strength of aluminum wire samples as a function of the amount of A206/1 wt % γAl$_2$O$_3$ added.

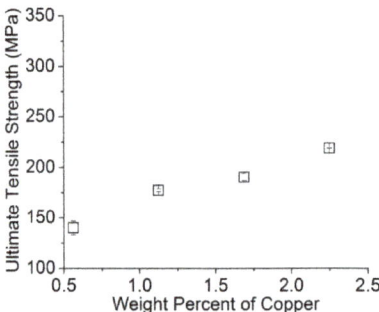

Figure 5. Average ultimate tensile strength of aluminum wire samples as a function of the amount of copper added.

3.2.2. Electrical Conductivity Measurements

The wires treated with the A206/1 wt % γAl$_2$O$_3$ nanocomposite and copper displayed lower electrical conductivity than aluminum at 25 °C, for alumina and copper concentrations ranging from 0.125 to 0.5 wt % and 0.562 to 2.250 wt %, respectively, as observed in Figures 6 and 7.

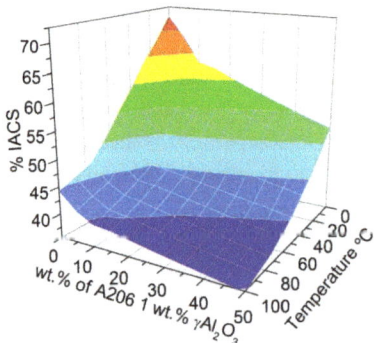

Figure 6. Effect of the amount of A206/1 wt % γAl$_2$O$_3$ nanocomposite added and the temperature on the electrical conductivity of aluminum wires (measured as a percent of International Annealed Copper Standard (IACS)).

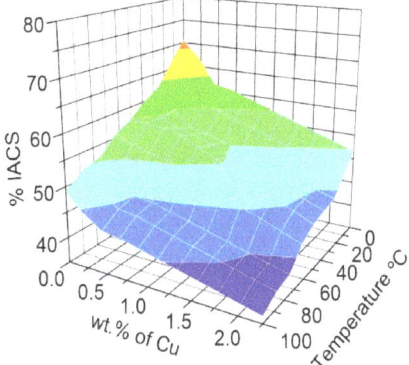

Figure 7. Effect of the amount of copper added and the temperature on the electrical conductivity of aluminum wires (measured as a percent of IACS).

3.2.3. Bending (Looping) Properties

To test the ability of a wire to be spooled without fracturing, the ASTM B 230/B 230M-07 standard can be used. This recommends looping the wire around its own diameter or about a mandrel [25]. This procedure reveals whether the wire is ductile enough to form a spool. As Figure 8 (A through C) demonstrates, the aluminum wires treated with A206/1 wt % γAl_2O_3 nanocomposites to 12.5, 25.0, and 37.5 weight percent did not display any fissure or crack. For the aluminum wire treated with 50.0 wt % of A206/1 wt % γAl_2O_3, two fractures became visible (Figure 8D). Thus, upon analyzing the bending properties and the macrograph of the wire fractures, one can conclude that increasing the A206/1 wt % γAl_2O_3 content of the samples caused ductility loss. This is an important finding that needs to be considered when manufacturing these types of wires.

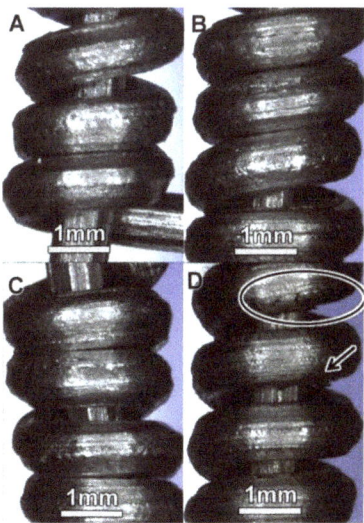

Figure 8. Bending (looping) test of aluminum wires treated with different weight percentage of A206/1 wt % γAl_2O_3 nanocomposite: (**A**) 12.5; (**B**) 25.0; (**C**) 37.5; and (**D**) 50.0.

3.2.4. Density of Wires

In Figure 9, it is apparent that the wire density increased slightly (no more than 1%) with the amount of A206/1 wt % γAl$_2$O$_3$ added to the melt. As expected, the alumina and copper content did affect the density of the wires; this is because the density of the alumina and copper were ~47% and ~231% higher than aluminum, respectively [24].

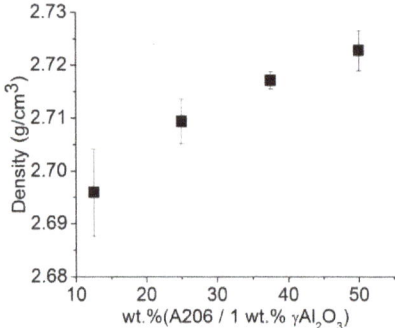

Figure 9. Measured density of aluminum wire samples as a function of the amount of A206/1 wt % γAl$_2$O$_3$ added.

3.2.5. Fractographic Study of Wires

Figure 10 shows the scanning electron images of the fractures after the tensile tests of the different wires studied. The public domain ImageJ image analysis software was used to measure the area percent of brittle and ductile fractures. The original images were binarized (black and white) where the brittle area was presented as black and the ductile as white. Then, on the calibrated black and white image, the percentage of brittle and ductile fractures was measured. This procedure was developed and used in previous work [26]. Brittle area measurements (in percent) are shown in Figure 11 where a higher amount of A206/1 wt % γAl$_2$O$_3$ added yielded more brittleness of the treated wires.

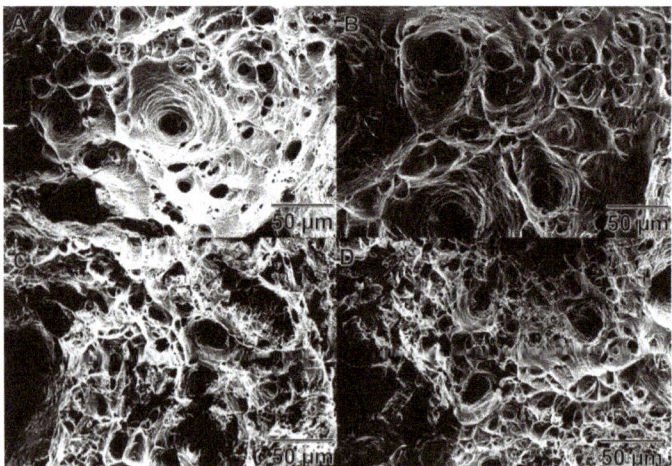

Figure 10. Secondary electron images of the tensile fractures of aluminum wires: (**A**) wires with 12.5 wt % (A206/1 wt % γAl$_2$O$_3$); (**B**) wires with 25.0 wt % (A206/1 wt % γAl$_2$O$_3$); (**C**) wires with 37.5 wt % (A206/1 wt % γAl$_2$O$_3$); and (**D**) wires with 50.0 wt % (A206/1 wt % γAl$_2$O$_3$).

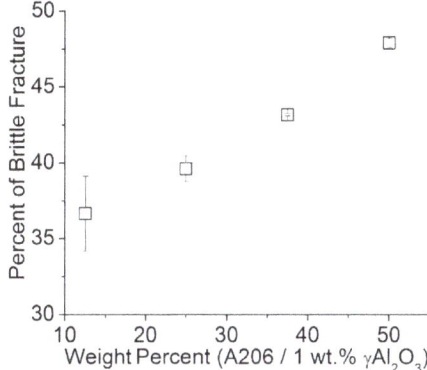

Figure 11. Percent of brittle fracture area in aluminum wires.

3.2.6. Wires Thermal Analysis

A differential thermal analysis apparatus allowed determining the onsets of melting (upon heating) and solidification (upon cooling) of the Al-A206/1 wt % γAl$_2$O$_3$ wire samples. The results are displayed in Figure 12 whereas Figure 13 shows the melting and solidification onsets as a function of the amount of Cu. By combining both graphs, it is apparent that the concentration of copper decreased the melting point of all wire samples, as expected. In effect, this is the natural behavior of aluminum–copper binary alloys in which the addition of copper decreases the liquidus temperature of a given Al–Cu alloy in the aluminum-rich region [27]. In addition, one could observe that the addition of Al$_2$O$_3$ particles maintained the initial solidification temperature almost constant and decreased the melting onset temperature of the wires by approximately 5 °C. The almost constant solidification onset temperature can be explained by heterogeneous nucleation of aluminum grains when the nanoparticles are likely acting as catalytic substrates for such nucleation events. While aluminum heterogeneous nucleation has been widely studied previously using differential thermal analysis [28–30], the use of Al$_2$O$_3$ particles as nucleation agents was, for instance, discussed by L. Yang [30]. The catalytic potency of the alumina nanoparticles as nucleants was more marked in the Cu-containing alloys. In effect, dissolved copper lowers the liquidus line, which in this case was prevented by the presence of the nanoparticles. Thus, they favored early nucleation events upon cooling.

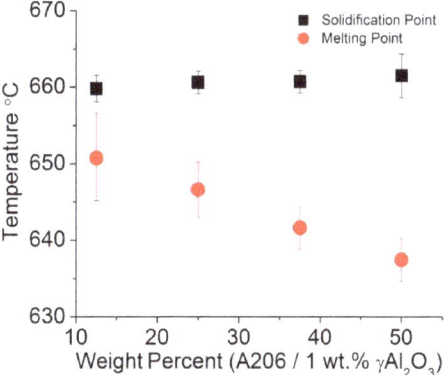

Figure 12. Melting and solidification onset temperatures of aluminum wire samples as measured in a differential thermal analysis apparatus.

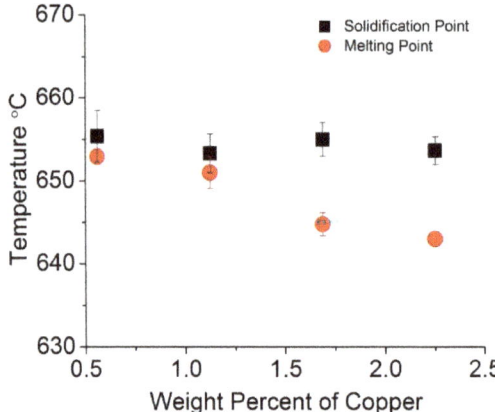

Figure 13. Melting and solidification onset temperatures measured in the Al–Cu wire samples using a differential thermal analyzer.

4. Discussion

To study the statistical significance of the effects of both alumina and copper additions (as mentioned, in the form of A206/1 wt % γAl_2O_3) on the ultimate tensile strength, electrical conductivity, and melting temperatures of the wires, we carried out a multiple linear regression analysis. The nomenclature used in the equations is as follows: % IACS = percent of International Annealed Copper Standard; T = temperature of the wires in degrees Celsius (measured simultaneously with the electrical conductivity using the four-point probe); UTS = ultimate tensile strength (MPa); % Cu = weight percent of copper, % γAl_2O_3 = weight percent of γAl_2O_3, and MP = measured melting point.

The analysis of variance (ANOVA) presented in Tables 3–5 for the three models provides the ensuing fitted parameters and resulting p-values. Equation (1) describes the ultimate tensile strength as a function of the alumina and copper levels; again, the resulting p-values for the wt % γAl_2O_3 and wt % Cu variables are nil. This indicates that the alumina and copper amounts were very effective in strengthening the wires.

$$UTS = 70.12 + 30.76 \cdot wt\% \, Cu + 595.99 \cdot wt\% \, Al_2O_3 - 143.89 \cdot wt\% \, Cu \cdot wt\% \, Al_2O_3 \qquad (1)$$

Table 3. Analysis of variance (ANOVA) of the model in Equation (1).

Parameter	Value	Standard Error of the Coefficient	p-Value
Constant	70.124	8.319	0.000
wt % Al_2O_3	595.989	62.778	0.000
wt % Cu	30.763	5.762	0.000
wt % Cu·wt % Al_2O_3	−143.890	32.424	0.001

The regression model in Equation (2) describes how the electrical conductivity of the wires varies as a function of the γAl_2O_3 and copper amounts, as well as temperature; the p-value is zero for all the parameters. This further corroborates that increasing the alumina and copper lowered the conductivity of the wires. In the case of γAl_2O_3, this can be readily explained by the high electric resistivity of alumina at room temperature, which is estimated to exceed $10^{12} \, \Omega \cdot m$ [22]. The decrease in conductivity of the wire when copper was added is explained by S. Aksöz [31].

$$\% \, IACS = 64.62 - 26.96 \cdot wt\% \, Al_2O_3 - 7.46 \cdot wt\% \, Cu - 0.13 \cdot T + 8.84 \cdot wt\% \, Cu \cdot wt\% \, Al_2O_3 \qquad (2)$$

Table 4. ANOVA of the model in Equation (2).

Parameter	Value	Standard Error of the Coefficient	p-Value
Constant	64.624	1.387	0.000
wt % Al$_2$O$_3$	−26.958	9.301	0.000
wt % Cu	−7.456	0.874	0.000
Temperature	−0.125	0.009	0.000
wt % Cu·wt % Al$_2$O$_3$	8.837	4.787	0.072

In Equation (3), the model describes the behavior of the melting point; the p-values are zero for wt % γAl$_2$O$_3$ and wt % Cu, which indicate that alumina and copper decreased significantly the melting points of the wires. The R^2 values for the resulting regression models (i.e., electrical conductivity, ultimate tensile strength, and melting point) are high: 97.10%, 86.95%, and 94.11%, respectively. This is an important finding if these wires are intended for fillers in TIG welding of aluminum parts: the strength of the welded joint can be improved with the addition of alumina to the filler. At the same time, as the melting point of the wire filler becomes lower, less energy would be required to melt the wire. This advantageous fact is further buttressed by a higher electrical resistivity of the wire (due to the alumina and copper added). In summary, combining both properties, i.e., lower melting points and higher resistivity, one can expect to decrease the energy needed to melt the filler when alumina nanoparticles are present. Nonetheless, one must take into consideration that the content of A206/1 wt % γAl$_2$O$_3$ must be less than 50% in order to form a sound spool (no cracks) of these novel wires.

$$MP = 656.22 - 11.32 \cdot wt \% \, Al_2O_3 - 5.99 \cdot wt \% \, Cu \quad (3)$$

Table 5. ANOVA of the model in Equation (3).

Parameter	Value	Standard Error of the Coefficient	p-Value
Constant	656.221	0.66204	0.000
wt % Al$_2$O$_3$	−11.315	1.57952	0.000
wt % Cu	−5.987	0.46425	0.000

5. Conclusions

The experimental results allow several conclusions:

- The γAl$_2$O$_3$ nanoparticles can be successfully added to molten A206 to fabricate an A206/1 wt % γAl$_2$O$_3$ nanocomposite by semi-solid mixing and ultrasonic processing.
- The ultimate tensile strength of the wires can be increased by increasing the amount of γAl$_2$O$_3$ nanoparticles and Cu added to the aluminum melt.
- Increasing the levels of γAl$_2$O$_3$ nanoparticles and Cu lowers the electrical conductivity and melting point of the wires.
- A fractography study revealed that an increment of A206/1 wt % γAl$_2$O$_3$ nanocomposite in the aluminum matrix leads to more brittleness of the wires.
- All those results are also corroborated via statistical analysis. They evince the feasibility of using this new material as a filler in aluminum welding.

Acknowledgments: The authors would like to thank the Materials Research Laboratory technician, Boris Rentería, and the former undergraduate students Alexandra Padilla and Grace Rodríguez for their assistance in completing this project. This material is based upon work supported by the US National Science Foundation under Grants HRD 0833112 and 1345156 (CREST program). The tensile test machine was acquired through a grant provided by the Solid Waste Management Authority of Puerto Rico.

Author Contributions: Xiaochun Li and Hongseok Choi manufactured the master A206/alumina nanocomposites and participated in the preparation of this manuscript. David Florián-Algarín and Raúl Marrero manufactured the wires, analyzed and interpreted the data obtained from the characterization of those wires and wrote a significant portion of the manuscript. Oscar Marcelo Suárez leads the Nanotechnology Center hosting this research and contributed also to the manuscript preparation.

Conflicts of Interest: The authors declare no conflict of interest.

References

1. Zhang, Z.D.; Liu, L.M.; Song, G. Welding characteristics of AZ31B magnesium alloy using DC-PMIG welding. *Trans. Nonferr. Met. Soc. China Engl. Ed.* **2013**, *23*, 315–322. [CrossRef]
2. Gao, M.; Tang, H.-G.; Chen, X.-F.; Zeng, X.-Y. High power fiber laser arc hybrid welding of AZ31B magnesium alloy. *Mater. Des.* **2012**, *42*, 46–54. [CrossRef]
3. Seshagiri, P.C.; Nair, B.S.; Reddy, G.M.; Rao, K.S.; Bhattacharya, S.S.; Rao, K.P. Improvement of mechanical properties of aluminum–copper alloy (AA2219) GTA welds by Sc addition. *Sci. Technol. Weld. Join.* **2008**, *13*, 146–158. [CrossRef]
4. Nie, J.F. Preface to viewpoint set on: Phase transformations and deformation in magnesium alloys. *Scr. Mater.* **2003**, *48*, 981–984. [CrossRef]
5. Jun, J.-H. Microstructure and Damping Capacity of Mg_2Si/Mg-Al-Si-(Bi) Composites. *Mater. Trans.* **2012**, *53*, 2064–2066. [CrossRef]
6. Fauzi, M.N.A.; Uday, M.B.; Zuhailawati, H.; Ismail, A.B. Microstructure and mechanical properties of alumina-6061 aluminum alloy joined by friction welding. *Mater. Des.* **2010**, *31*, 670–676. [CrossRef]
7. Qin, Q.D.; Huang, B.W.; Li, W.; Zeng, Z.Y. Preparation and wear resistance of aluminum composites reinforced with in situ formed TiO/Al_2O_3. *J. Mater. Eng. Perform.* **2016**, *25*, 2029–2036. [CrossRef]
8. Hoseini, M.; Meratian, M. Tensile properties of in-situ aluminum–alumina composites. *Mater. Lett.* **2005**, *59*, 3414–3418. [CrossRef]
9. Jiang, X.; Wang, N.; Zhu, D. Friction and wear properties of in-situ synthesized Al_2O_3 reinforced aluminum composites. *Trans. Nonferr. Met. Soc. China* **2014**, *24*, 2352–2358. [CrossRef]
10. Belhouchet, H.; Garnier, V.; Fantozzi, G.; Franc, J. Control of the γ-alumina to α-alumina phase transformation for an optimized alumina densification. *Bol. Soc. Esp. Cerám. Vidrio* **2016**, *56*, 4–11.
11. Yalamaç, E.; Trapani, A.; Akkurt, S. Sintering and microstructural investigation of gamma e alpha alumina powders. *Eng. Sci. Technol. Int. J.* **2014**, *17*, 2–7. [CrossRef]
12. Dorri, A.; Omrani, E.; Lopez, H.; Zhou, L.; Sohn, Y.; Rohatgi, P.K. Strengthening in hybrid alumina-titanium diboride aluminum matrix composites synthesized by ultrasonic assisted reactive mechanical mixing. *Mater. Sci. Eng. A* **2017**, *702*, 312–321. [CrossRef]
13. *ASM Handbook Properties and Selection: Nonferrous Alloys and Special-Purpose Materials*, 10th ed.; ASM International: Geauga County, OH, USA, 1990; Volume 2, pp. 78–79.
14. Sreekumar, V.M.; Babu, N.H.; Eskin, D.G. Potential of an Al-Ti-$MgAl_2O_4$ master alloy and ultrasonic cavitation in the grain refinement of a cast aluminum alloy. *Metall. Mater. Trans. B* **2017**, *48*, 208–219. [CrossRef]
15. Choi, H.; Cho, W.H.; Konishi, H.; Kou, S.; Li, X. Nanoparticle-induced superior hot tearing resistance of A206 alloy. *Metall. Mater. Trans. A* **2013**, *44*, 1897–1907. [CrossRef]
16. Wu, J.; Zhou, S.; Li, X. Ultrasonic attenuation based inspection method for scale-up production of A206-Al_2O_3 metal matrix nanocomposites. *J. Manuf. Sci. Eng.* **2014**, *137*, 1–10.
17. Chen, L.Y.; Weiss, D.; Morrow, J.; Xu, J.Q.; Li, X.C. A novel manufacturing route for production of high-performance metal matrix nanocomposites. *Manuf. Lett.* **2013**, *1*, 62 65. [CrossRef]
18. Florián-Algarín, D.; Marrero, R.; Padilla, A.; Suárez, O.M. Strengthening of Al and Al- Mg alloy wires by melt inoculation with Al/MgB_2 nanocomposite. *J. Mech. Behav. Mater.* **2015**, *24*, 207–212. [CrossRef]
19. Florián-Algarín, D.; Padilla, A.; López, N.N.; Suárez, O.M. Fabrication of aluminum wires treated with nanocomposite pellets. *Sci. Eng. Compos. Mater.* **2014**, *22*, 485–490. [CrossRef]
20. B557-06 ASTM. *Standard Test Methods for Tension Testing Wrought and Cast Aluminum-and Magnesium-Alloy Products*; ASM International: Geauga County, OH, USA, 2010; pp. 1–15.
21. Suárez, O.M.; Stone, D.S.; Kailhofer, C.J. Measurement of electrical resistivity in metals and alloys using a commercial data acquisition software. *J. Mater. Educ.* **1999**, *20*, 341–356.

22. Stratton, S.W. *U.S. National Bureau of Standards Copper Wire Tattles*, 3rd ed.; Washington Government Printing Office: Washington, DC, USA, 1914.
23. Dieringa, H. Properties of magnesium alloys reinforced with nanoparticles and carbon nanotubes: A review. *J. Mater. Sci.* **2010**, *46*, 289–306. [CrossRef]
24. *ASM Handbook Properties and Selection: Nonferrous Alloys and Special-Purpose Materials*, 10th ed.; ASM International: Geauga County, OH, USA, 1990; Volume 2, p. 17.
25. *ASTM B230/B230M. Standard Specification for Aluminum 1350—H19 Wire for Electrical Purposes 1*; ASM International: Geauga County, OH, USA, 1975; pp. 1–4.
26. Corchado, M.; Reyes, F.; Suárez, O.M. Effects of AlB$_2$ Particles and zinc on the absorbed impact energy of gravity cast aluminum matrix composites. *JOM* **2014**, *66*, 926–934. [CrossRef]
27. Yan, X.-Y.; Chang, Y.; Xie, F.-Y.; Chen, S.-L.; Zhang, F.; Daniel, S. Calculated phase diagrams of aluminum alloys from binary Al–Cu to multicomponent commercial alloys. *J. Alloys Compd.* **2001**, *320*, 151–160. [CrossRef]
28. Suarez, O.M.; Perepezko, J.H. Microstructural observation of active nucleants in Al-Ti-B master alloys. *Light Met.* **1992**, *3*, 851–859.
29. Suárez, O.M. A study of the role of diborides in the heterogeneous nucleation of aluminum. *Rev. Met.* **2004**, *40*, 173–181.
30. Yang, L.; Xia, M.; Li, J.G. Epitaxial growth in heterogeneous nucleation of pure aluminum. *Mater. Lett.* **2014**, *132*, 52–54. [CrossRef]
31. Aksöz, S.; Ocak, Y.; Maraşlı, N.; Çadirli, E.; Kaya, H.; Böyük, U. Dependency of the thermal and electrical conductivity on the temperature and composition of Cu in the Al based Al–Cu alloys. *Exp. Therm. Fluid Sci.* **2010**, *34*, 1507–1516. [CrossRef]

© 2018 by the authors. Licensee MDPI, Basel, Switzerland. This article is an open access article distributed under the terms and conditions of the Creative Commons Attribution (CC BY) license (http://creativecommons.org/licenses/by/4.0/).

Article

The Effect of Zn Content on the Mechanical Properties of Mg-4Nd-xZn Alloys (x = 0, 3, 5 and 8 wt.%)

Serge Gavras [1,*], Ricardo H. Buzolin [2], Tungky Subroto [1], Andreas Stark [1] and Domonkos Tolnai [1]

1. Institute of Materials Research, Helmholtz-Zentrum Geesthacht, Max-Planck-Strasse 1, D 21502 Geesthacht, Germany; tungky.subroto@hzg.de (T.S.); andreas.stark@hzg.de (A.S.); domonkos.tolnai@hzg.de (D.T.)
2. Institute of Materials Science, Joining and Forming, Graz University of Technology, A8010 Graz, Austria; ricardo.buzolin@tugraz.at
* Correspondence: sarkis.gavras@hzg.de; Tel.: +49-4152-87-1912

Received: 30 April 2018; Accepted: 25 June 2018; Published: 28 June 2018

Abstract: The mechanical properties of as-cast Mg-4Nd-xZn (x = 0, 3, 5 or 8 wt.%) alloys were investigated both in situ and ex situ in as-cast and solution-treated conditions. The additions of 3 or 5 wt.% Zn in the base Mg-4Nd alloy did not improve yield strength in comparison to the binary Mg-4Nd alloy. Mechanical properties were shown to improve only with the relatively high concentration of 8 wt.% Zn to Mg-4Nd. The change in intermetallic morphology from a continuous intermetallic to a lamella-like intermetallic was the primary reason for the decreased mechanical properties in Mg-4Nd-3Zn and Mg-4Nd-5Zn compared with Mg-4Nd and Mg-4Nd-8Zn. The dissolution of intermetallic at grain boundaries following heat treatment further indicated the importance of grain boundary reinforcement as shown in both in situ and ex situ compression testing. Azimuthal angle-time plots indicated little grain rotation most noticeably in Mg-4Nd, which also indicated the influence of a strong intermetallic network along the grain boundaries.

Keywords: magnesium alloys; zinc addition; neodymium; Mg-Nd-Zn alloys; deformation behaviour; in situ synchrotron radiation diffraction

1. Introduction

The growing demand for increased efficiency in the transportation sectors has directed attention to lightweight solutions. Magnesium, as one of the lightest metals [1] was the focus of numerous pieces of research in recent years with the goal to overcome the effects of poor absolute mechanical properties and corrosion resistance [2]. This prior research lead not only to the improvement of the Mg alloys' performance, but it was also shown that controlled corrosion allows for the use of certain Mg alloys as degradable implants in the medical sector [3].

Alloying Mg with Zn results in a combination of enhanced strength and ductility [4]. Owing to this performance, the alloy ZK60 (Mg-6.0Zn-0.6Zr (wt.%)) became one of the highest strength commercial wrought Mg alloy. Thus, the Mg-Zn alloys served as a foundation for the development of low cost Mg alloys and different elemental additions to this system has been investigated in order to develop the mechanical property profile through engineering of the grain boundary phases [5–9]. These investigations report further enhancement of elevated temperature strength and ductility [10,11] by the addition of Rare earth (RE) elements to the Mg-Zn-Zr system [12]. Neodymium having a relatively low solid solubility in Mg (3.6 wt.% at 549 °C [13]) is an ideal RE element because high concentrations are not needed to produce second phase particles in order to further improve the elevated temperature strength [14,15]. Furthermore, Nd is not toxic therefore the Mg-Nd-Zn system is under investigations for bio absorbable implant materials [16].

The evolution of synchrotron radiation based high energy X-ray and neutron diffraction advanced in situ characterisation methods [17,18]. The transmission geometry allows investigating bulk materials undergoing thermo-mechanical loading and to follow the dynamic microstructural processes occurring during processing [19]. The continuously acquired diffraction patterns and their evolution provide information on the grain structure and its changes, texture evolution, strain and strain anisotropy, which can be correlated with dislocation slip, twinning, recrystallization and recovery [20,21].

The tensile properties of as-cast Mg-4Nd and Mg-4Nd-8Zn at room temperature and 200 °C were investigated previously by Gavras et al. [22]. This current work builds on this by examining the influences of different Zn concentrations and solution treatment on Mg-Nd and Mg-Nd-Zn alloys.

2. Materials and Methods

A Mg-4Nd binary alloy was used as a base alloy and compared with ternary Mg-4Nd-xZn alloys with additions of 3, 5 and 8 wt.% of Zn in as-cast and solution treated conditions. The alloys were prepared by permanent mould indirect chill casting [23], an electric resistance furnace was used to melt the Mg under protective atmosphere of 2 vol.% SF_6 and Ar. Zn and Nd were added as pure elements. After mixing, the melt was held at 720 °C for 10 min, then poured into a steel mould preheated to 660 °C. After holding at this temperature for 5 min, the mould was quenched in room temperature water at a rate of 10 mm s^{-1} until the top of the melt was in line with the cooling water level. Solution treatments were performed for 24 h at 525 °C for the Mg-4Nd, 480 °C for the Mg-4Nd-3Zn, 440 °C for the Mg-4Nd-5Nd and 420 °C for the Mg-4Nd-8Zn to prevent the partial melting of the intermetallics present in the different alloys.

For metallographic characterisation, as-cast and deformed samples were mounted in epoxy and ground using SiC paper and then polished using OPS solution. The microstructures for optical microscopy (OM) were etched with acetic-picral solution. The specimens for electron backscattered diffraction studies (EBSD) were not etched but washed immediately before the measurements with 0.5 vol.% nitric acid in ethanol solution for 5 s.

The optical microscopic analysis was performed using a Leica DMI 5000 light optical microscope (Wetzlar, Germany). Grain size measurements were obtained from optical micrographs using the line intercept method over an area of the sample covering approximately 100 grains. The scanning electron microscopy investigation was performed using a Tescan Vega3 SEM (Brno, Czech Republic) and a Zeiss FEG-SEM Ultra 55 (Oberkochen, Germany) attached with a Hikari detector (EDAX, Weiterstadt, Germany) and a TSL-OIM software (EDAX, Weiterstadt, Germany) package for Electron Backscatter Diffraction (EBSD) analysis. The EBSD measurements on compressed samples were conducted on an area of 700 μm × 700 μm with step size of 1 μm close to the centre of the deformed specimen to ensure a similar area to that measured during diffraction analysis. All microstructures of deformed samples are presented so that the compression direction is parallel to the horizontal direction.

The tensile and compression tests were performed on a Zwick Z050 universal testing machine (Zwick/Roell, Ulm, Germany). A maximum load of 50 kN was used to test samples at room temperature and at 200 °C, using a strain rate of $10^{-3} \cdot s^{-1}$. The tensile and compression tests were performed using the standard DIN 50125 [24]. with a minimum of five samples per condition and per alloy. Tensile and compression samples were cut from ingots using electron discharge machining (EDM). The dimensions of the tensile samples were 60 mm in total length with a gauge length of 35 mm and gauge diameters of 9.8 mm. The compression samples had a length of 15 mm with a diameter of 10 mm.

For the in situ compression experiments, cylindrical specimens were machined from cast ingots and solution treated specimens with a diameter of 5 mm and length of 10 mm. The in situ synchrotron radiation diffraction was performed at the P07 beamline of Petra III, DESY (Deutsches Elektronen-Synchrotron, Hamburg, Germany). A monochromatic beam with the energy of 100 keV (λ = 0.0124 nm) and with a cross section of 1 × 1 mm^2 was used. Diffraction patterns were recorded with a PerkinElmer 1622 flat panel detector (Baesweiler, Germany) with a pixel size of (200 μm^2)

which was placed at a sample-to-detector distance of 1603 mm from the specimen (calibrated with a LaB_6 standard powder sample). The acquisition time for each image was 1 s. The specimens were placed in the chamber of a dilatometer DIL 805A/D (TA Instruments, New Castle, DE, USA), combined with a modified heating induction coil in order for the beam to pass only through the sample [25]. The specimens were compressed at room temperature and at 200 °C. For the tests at 200 °C, the specimens were heated to the test temperature at a rate of 30 K·s^{-1} and held for 3 min before the compression started to ensure temperature homogeneity. The specimens were compressed with an initial strain rate of 1.0×10^{-3} s^{-1}. The tests were terminated at a strain of 0.1. The morphology of the Debye-Scherrer rings was then analysed using the Fit2D® software (ESRF, Grenoble, France) and converted into azimuthal-angle time (AT) plots by using the ImageJ® software package (NIH, Bethesda, MA, USA).

3. Results

3.1. Metallography

The actual alloy compositions measured with spark analyser and X-ray fluorescence spectroscopy are listed in Table 1.

Table 1. Chemical compositions of the alloys.

Alloy (wt.%)	Nd wt.% (XRF)	Zn wt.% (Spark Analyzer)
Mg-4Nd	4.20	–
Mg-4Nd-3Zn	4.35	3.20
Mg-4Nd-5Zn	4.20	5.20
Mg-4Nd-8Zn	4.34	8.00

The microstructure and mechanical properties of Mg-4Nd, Mg-4Nd-3Zn, Mg-4Nd-5Zn and Mg-4Nd-8Zn were compared in as-cast and solution treated conditions. Solution treatments were chosen in order to get the highest concentrations of alloying additions into solid solution without causing the intermetallic phases present at grain boundaries to melt. The addition and then increasing concentration of Zn to the base Mg-4Nd alloys decreases the average grain size of the alloys (Table 2).

Table 2. Average grain size of as-cast and solution treated alloys.

Alloy (wt.%)	Grain Size in As-Cast Condition (mm) ± SD	Grain Size in Solution Treated Condition (mm) ± SD
Mg-4Nd	0.99 ± 0.14	1.13 ± 0.07
Mg-4Nd-3Zn	0.55 ± 0.03	0.76 ± 0.04
Mg-4Nd-5Zn	0.36 ± 0.02	0.38 ± 0.01
Mg-4Nd-8Zn	0.20 ± 0.02	0.14 ± 0.02

The addition of Zn to the binary Mg-4Nd alloy results in a change to the morphology of the intermetallic phase present at grain boundaries. Figure 1 is used to illustrate the change in intermetallic morphology with increasing Zn additions. Due to the relatively high concentrations of Nd and Zn additions to Mg, not all of the intermetallic dissolved into solid solution following solution treatments.

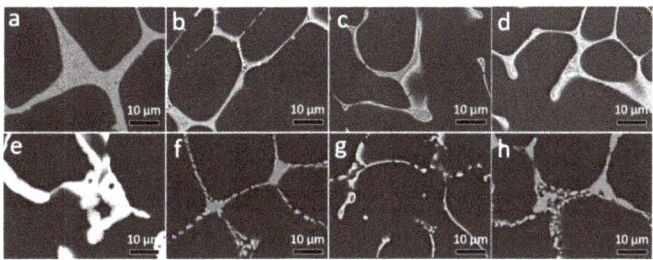

Figure 1. BSE SEM micrographs of as-cast (**a**) Mg-4Nd, (**b**) Mg-4Nd-3Zn, (**c**) Mg-4Nd-5Zn and (**d**) Mg-4Nd-8Zn and solution treated (**e**) Mg-4Nd (525 °C for 24 h), (**f**) Mg-4Nd-3Zn (480 °C for 24 h), (**g**) Mg-4Nd-5Zn (440 °C for 24 h) and (**h**) Mg-4Nd-8Zn (420 °C for 24 h).

3.2. Mechanical Properties

The binary Mg-4Nd alloy has superior tensile properties compared with the ternary Mg-4Nd-3Zn and Mg-4Nd-5Zn alloys (Figure 2). It is only when the relatively high concentration of 8 wt.% Zn is added to the Mg-4Nd that the ternary alloy properties improve. However, this is still not a significant improvement to tensile properties. In the as-cast condition, the addition of 8 wt.% Zn to Mg-4Nd, at best, only marginally improves the 0.2% proof stress (PS) and ultimate tensile strength (UTS) compared with Mg-4Nd. In the solution treated condition, however, the Mg-4Nd-8Zn performs significantly better than Mg-4Nd particularly with respect to the UTS (Table 3). The Mg-4Nd-5Zn alloy, in general, has significantly poorer 0.2% PS and UTS compared to the other alloys tested (Table 3). The elongation to failure properties are poor for all alloys in both as-cast and solution treated conditions. Elongation values do not exceed 5% at room temperature (Table 3). It is only at 200 °C when the Mg-4Nd-8Zn alloy in as-cast or solution treated conditions has an elongation to failure value greater than 10%.

Following solution treatment at 520 °C for 24 h the binary Mg-4Nd alloy has poorer room temperature and 200 °C tensile properties compared with both solution treated Mg-4Nd-3Zn (480 °C for 24 h) and Mg-4Nd-8Zn (420 °C for 24 h), Figure 2c,d. The Mg-4Nd-5Zn alloy continues to have the lowest 0.2% PS and UTS of all alloys tested (Table 3). It is important to note that solution treatment did not improve tensile properties for any of the alloys tested with exception to Mg-4Nd-8Zn.

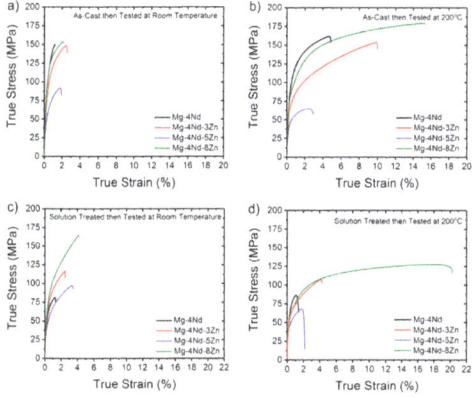

Figure 2. Representative tensile true stress—true strain curves for the as-cast and solution treated alloys in conditions (**a**) as-cast at room temperature, (**b**) as-cast at 200 °C, (**c**) solution treated at room temperature and (**d**) solution treated at 200 °C.

Table 3. Ex situ tensile properties of alloys tested at room temperature or 200 °C in as-cast or solution treated conditions.

Alloy (wt.%)	Ave 0.2% PS ± SD (MPa)	Ave UTS ± SD (MPa)	Ave Elong ± SD (%)
	As-Cast Tested at Room Temperature		
Mg-4Nd	103.4 ± 1.8	147.7 ± 17.0	1.2 ± 0.3
Mg-4Nd-3Zn	81.7 ± 2.1	143.6 ± 6.6	2.5 ± 0.3
Mg-4Nd-5Zn	50.0 ± 1.5	84.9 ± 10.1	1.7 ± 0.7
Mg-4Nd-8Zn	105.1 ± 1.9	151.4 ± 2.1	2.0 ± 0.1
	Solution Treated Tested at Room Temperature		
Mg-4Nd	53.6 ± 4.5	80.2 ± 28.5	2.0 ± 1.0
Mg-4Nd-3Zn	66.7 ± 11.7	113.0 ± 20.1	2.5 ± 0.7
Mg-4Nd-5Zn	46.1 ± 10.9	107.7 ± 31.1	3.2 ± 0.7
Mg-4Nd-8Zn	71.1 ± 3.1	163.9 ± 27.7	4.3 ± 1.7
	As-Cast Tested at 200 °C		
Mg-4Nd	96.8 ± 2.1	169.3 ± 12.1	4.9 ± 1.3
Mg-4Nd-3Zn	69.7 ± 2.7	152.8 ± 3.4	9.9 ± 1.4
Mg-4Nd-5Zn	44.9 ± 1.0	66.1 ± 3.2	3.2 ± 0.8
Mg-4Nd-8Zn	83.9 ± 3.2	177.7 ± 4.7	15.2 ± 0.5
	Solution Treated Tested at 200 °C		
Mg-4Nd	68.5 ± 15.5	83.4 ± 4.7	1.6 ± 0.3
Mg-4Nd-3Zn	62.3 ± 9.5	97.1 ± 8.6	3.7 ± 1.0
Mg-4Nd-5Zn	43.8 ± 1.3	66.9 ± 3.1	1.9 ± 0.4
Mg-4Nd-8Zn	61.4 ± 1.3	125.1 ± 2.6	17.8 ± 6.8

Under compression testing conditions, the alloys experienced approximately twice the amount of ultimate compressive strength (UCS) in comparison to UTS at room temperature in the as-cast condition (Table 4). The compression to failure experienced by the alloys is also greater than the elongation to fracture. The most notable change in the trend of property-to-alloying addition relation is that the Mg-4Nd-5Nd does not have the poorest compression properties in the as-cast condition at room temperature (Figure 3a). This is in contrast to the Mg-4Nd-5Zn alloy under tension at room temperature. There is no significant change to the alloy properties under compression at 200 °C in the as-cast condition to the room temperature compression properties. However, the binary Mg-4Nd alloy has the lowest UCS (Figure 3c) compared to the other alloys tested. This is once again different to the tensile properties of room temperature (RT) as-cast alloys.

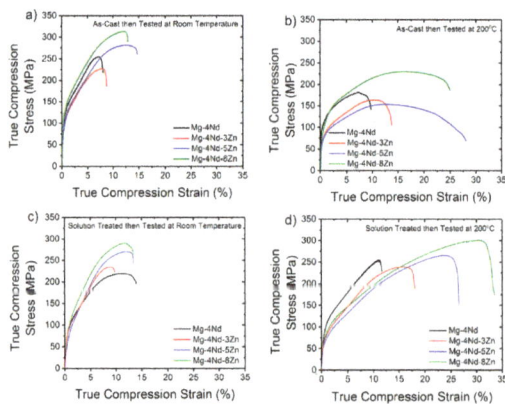

Figure 3. Compression true stress-strain curves for the as-cast and solution treated alloys in conditions (**a**) as-cast at room temperature, (**b**) as-cast at 200 °C, (**c**) solution treated at room temperature and (**d**) solution treated at 200 °C.

Similar to the as-cast condition, the alloys tested at room temperature and 200 °C following solution treatment experience higher UCS than UTS. In both tension and compression at RT for as-cast and solution treated conditions, Mg-4Nd-8Zn has the highest 0.2% PS and Mg-4Nd-5Zn has the lowest 0.2% PS compared to the other alloys tested in the same conditions.

Table 4. Ex-situ compression properties of alloys tested at room temperature or 200 °C in as-cast or solution treated conditions.

Alloy (wt.%)	Ave 0.2% PS ± SD (MPa)	Ave UCS ± SD (MPa)	Ave Comp. ± SD (%)
As-Cast Tested at Room Temperature			
Mg-4Nd	103.2 (±5.0)	257.4 (±11.3)	8.3 (±0.7)
Mg-4Nd-3Zn	87.9 (±2.8)	226.3 (±6.7)	9.0 (±0.9)
Mg-4Nd-5Zn	88.9 (±1.3)	283.1 (±5.0)	13.7 (±1.0)
Mg-4Nd-8Zn	115.0 (±1.6)	311.9 (±4.4)	13.0 (±1.7)
Solution Treated Tested at Room Temperature			
Mg-4Nd	87.1 (±2.4)	285.0 (±20.6)	14.4 (±2.1)
Mg-4Nd-3Zn	80.9 (±2.4)	285.1 (±6.2)	11.3 (±1.2)
Mg-4Nd-5Zn	69.0 (±5.1)	356.6 (±14.1)	13.9 (±1.2)
Mg-4Nd-8Zn	78.9 (±3.7)	375.4 (±3.7)	13.2 (±0.2)
As-Cast Tested at 200 °C			
Mg-4Nd	99.5 (±7.1)	198.8 (±13.1)	10.2 (±1.0)
Mg-4Nd-3Zn	68.8 (±3.2)	203.7 (±10.9)	12.3 (±1.1)
Mg-4Nd-5Zn	67.8 (±1.9)	218.7 (±9.4)	25.8 (±2.1)
Mg-4Nd-8Zn	89.7 (±0.4)	230.1 (±0.3)	26.4 (±1.0)
Solution Treated Tested at 200 °C			
Mg-4Nd	87.3 (±4.2)	258.3 (±5.5)	12.4 (±0.6)
Mg-4Nd-3Zn	67.2 (±1.4)	238.6 (±5.5)	18.0 (±2.0)
Mg-4Nd-5Zn	55.8 (±3.4)	265.9 (±8.6)	27.8 (±2.0)
Mg-4Nd-8Zn	69.3 (±0.5)	300.2 (±9.8)	33.9 (±1.5)

3.3. In Situ Compression Experiments

The true stress-true strain curves from the in situ compression tests are shown in Figure 4. Regarding the as-cast alloys tested at room temperature, a decrease in the maximum compressive stress was observed with the addition of Zn for the Mg-4Nd-3Zn and Mg-4Nd-5Zn. For the Mg-4Nd-8Zn the proof and maximum compressive stresses are comparable with the Mg-4Nd. For the tests performed at 200 °C, the Mg-4Nd exhibited the highest proof stress, followed by Mg-4Nd-8Zn. Mg-4Nd-5Zn and Mg-4Nd-3Zn exhibited comparable proof stresses. The maximum compressive stress was exhibited by Mg-4Nd-8Zn.

The in situ compression tests at room temperature of the T4 specimens show that Mg-4Nd exhibited the highest proof and maximum compression stresses, followed by the Mg-4Nd-8Zn (Table 5). The Mg-4Nd-3Zn and Mg-4Nd-5Zn alloys exhibited the lowest values. At 200 °C, the Mg-4Nd alloy also exhibited the highest proof and maximum compression stresses. The decrease in the compressive strength comparing to room temperature and 200 °C was more significant for the Mg-4Nd-5Zn, which exhibited the lowest proof and maximum compressive stresses at 200 °C. In both in situ and ex situ compression investigations the Mg-4Nd-5Zn alloy is shown to have the lowest 0.2% PS in comparison to the other alloys tested.

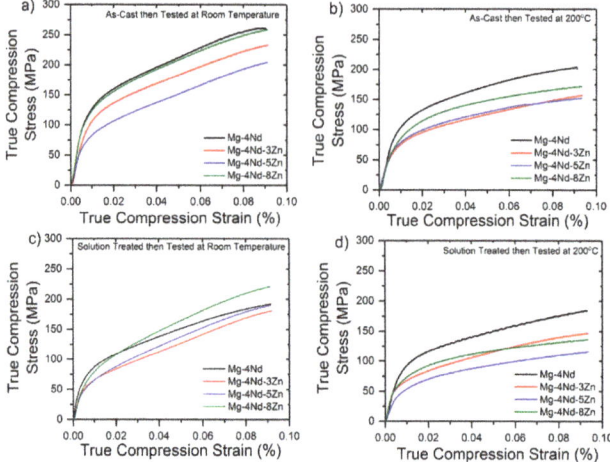

Figure 4. In situ compressive true stress-true strain curves for the as-cast and solution treated alloys in conditions (**a**) as-cast at room temperature, (**b**) as-cast at 200 °C, (**c**) solution treated at room temperature and (**d**) solution treated at 200 °C.

Table 5. In situ compression 0.2% proof stress of alloys tested at room temperature or 200 °C in as-cast or solution treated conditions.

	0.2% Proof Stress ± SD (MPa)			
Alloy (wt.%)	As-Cast at RT	As-Cast at 200 °C	Solution Treated at RT	Solution Treated at 200 °C
Mg-4Nd	106.8 (± 5.2)	80.4 (± 3.8)	63.2 (±1.5)	74.4 (±1.4)
Mg-4Nd-3Zn	88.2 (±2.9)	66.9 (±1.6)	56.2 (±2.9)	50.5 (±3.3)
Mg-4Nd-5Zn	69.2 (±1.5)	66.7 (±1.5)	50.2 (±1.4)	48.3 (±2.4)
Mg-4Nd-8Zn	100.3 (±4.4)	83.4 (±8.0)	63.5 (±0.1)	57.7 (±1.6)

Figure 5 shows the Azimuthal-angle time plots (AT)-plots for the as-cast alloys. Regarding the as-cast alloys, the AT-plots of the alloy compressed at 200 °C, the bending of the timelines was slightly more pronounced than at room temperature. Nevertheless, the bending of the timelines was very low for all alloys up to 10% compression, either at room temperature or 200 °C. The small grain rotation suggests that the microstructure is very stable, which can be explained by the presence of a rigid network of intermetallic compounds along the grain boundaries. The broadening of the timelines was more pronounced at 200 °C compared with room temperature, indicating that subgrain formation has occurred during deformation at this temperature. The as-cast alloy that exhibited the most stable microstructure, i.e., negligible grain rotation and not significant subgrain formation was the Mg-4Nd alloy. Discrete spots in the AT-plots were not observed after the beginning of deformation for any as-cast alloy neither deformed at room temperature nor at 200 °C. This indicates that discontinuous dynamic recrystallization did not play any role on the compression mechanisms of the investigated alloys.

Similar to the results for the as-cast alloys, the heat treated Mg-Nd-Zn alloys did not exhibit significant microstructural changes that are indicated in the AT-plots. The appearance of new timelines at the early stages of deformation is apparent for all alloys. The Mg-4Nd is the only alloy that also shows the appearance of timelines at late stages of deformation, especially for the $(10\bar{1}0)$ plane, suggesting intensive twinning during compression. Although a number of new spots in the AT-plots of the Mg-4Nd alloy deformed at 200 °C are visible, no strong evidence that discontinuous dynamic recrystallization played an important role on the deformation of the Mg-4Nd is observed in the microstructure.

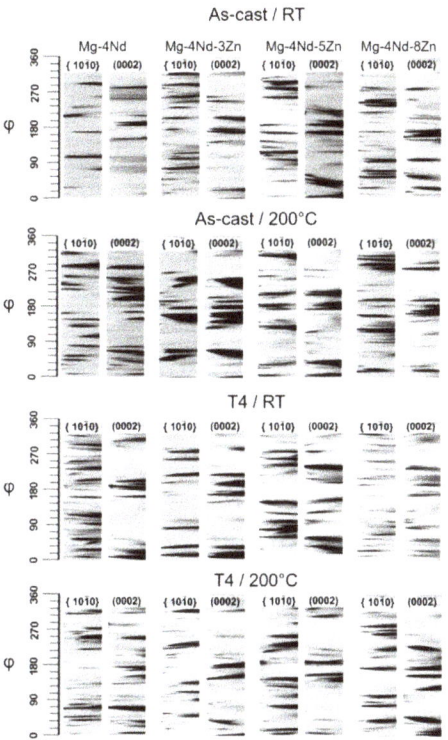

Figure 5. Azimuthal-angle time plots (AT) plots obtained during the in situ synchrotron radiation diffraction during compressive deformation at room temperature and 200 °C.

Figure 6 shows the IPF maps of the alloys for the as-cast alloys in situ compressed at room temperature Figure 6a,c,e,g and at 200 °C Figure 6b,d,f,h. For the alloys deformed at room temperature, fine twins are present along the grains that were in favourable orientation for twinning. A similar microstructure is observed for Mg-4Nd-3Zn, with slightly higher volume fraction of twins in comparison with Mg-4Nd. Mg-4Nd-5Zn alloy also shows a notably twinned microstructure. The thickness of the twins seems to be larger for the Mg-4Nd-5Zn, which suggests that there were regions with significant stress concentration during deformation in this alloy. The misorientation spread for the solution treated alloys compressed at 200 °C was significantly smaller compared with the alloys compressed at room temperature (Figure 7), suggesting that dynamic recovery played an important role on the deformation of the Mg-Nd-Zn alloys at this temperature. No evidence of recrystallised grains originating from discontinuous dynamic recrystallization was found. For Mg-4Nd some twins could be observed. Please define, if appropriate.

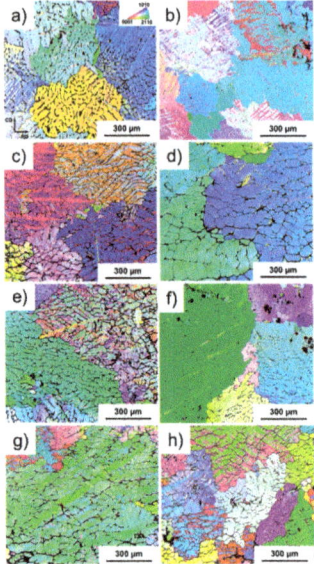

Figure 6. EBSD inverse pole figure maps of as-cast samples subjected to compressive strain at RT (**a,c,e,g**) and at 200 °C (**b,d,f,h**) of Mg-4Nd (**a,b**), Mg-4Nd-3Zn (**c,d**), Mg-4Nd-5Zn (**e,f**) and Mg-4Nd-8Zn (**g,h**).

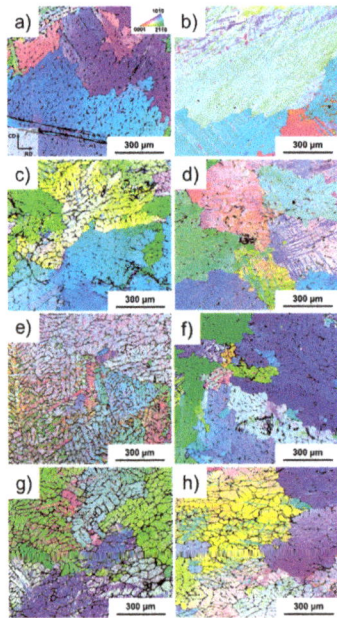

Figure 7. EBSD inverse pole figure maps of solution treated samples subjected to compressive strain at RT (**a,c,e,g**) and at 200 °C (**b,d,f,h**) of Mg-4Nd (**a,b**), Mg-4Nd-3Zn (**c,d**), Mg-4Nd-5Zn (**e,f**) and Mg-4Nd-8Zn (**g,h**).

4. Discussion

As shown previously, the intermetallic phase in the binary Mg-4Nd is $Mg_{12}Nd$ [26]. The addition of 3, 5 and 8 wt.% Zn to the base Mg-4Nd alloy results in two distinct intermetallic morphologies being present at grain boundaries, $Mg_{50}Nd_8Zn_{42}$ and $Mg_3(Nd, Zn)$ [22].

It is not immediately apparent as to why the Mg-4Nd-5Zn alloy generally performed more poorly than any of the other alloys investigated in this currently work. However, by investigation of the microstructure it can been seen that the intermetallic morphology of the Mg-4Nd is broad and continuous at grain boundaries (Figure 1a). The intermetallic morphology of Mg-4Nd-5Zn differs to that of Mg-4Nd (Figure 1). The Mg-4Nd-5Zn alloy has a less continuous and more lamellar-like appearance. The morphologies of the intermetallic present in Mg-4Nd-3Zn and Mg-4Nd-8Zn are more similar to that of Mg-4Nd-5Zn. This indicates that the continuous intermetallic phase ($Mg_{12}Nd$) has a significant strengthening effect on the Mg-4Nd alloy which was similarly shown in Mg-Nd and Mg-La-Nd alloys by Zhang et al. [27]. Particularly since the volume fraction of the $Mg_{12}Nd$ phase in Mg-4Nd is less than that of the intermetallic in the ternary alloys investigated (Table 2). However, with sufficient addition of Zn (namely Mg-4Nd-8Zn) the tensile and compressive properties of Mg-Nd-Zn is improved.

The dissolution of intermetallics at grain boundaries following solution treatment further indicates the importance of a strengthening intermetallic to obtain superior mechanical properties. The general reduction of 0.2% PS and ultimate tensile and compression strength of the solution treated alloys in comparison to as-cast alloys is also related to the reduction of grain boundary reinforcement. This indicates that although the intermetallic morphology present in the ternary alloys appears to be less able to withstand tensile or compressive loads, once a sufficient amount of grain boundary reinforcement is present, tensile or compressive properties improve.

In the AT plots each timeline corresponds to a coherent domain that is in Bragg position and therefore the diffraction occurs. In this case, when the timeline is a single line that indicates that the misorientation spread within the coherent domain is negligible [19]. A cast or recrystallized microstructure will exhibit discrete timelines. When the material starts to deform the microstructure must accommodate the strain. Geometrical conditions will give the most favourable deformation mechanism: basal glide or twinning. The disappearance of a timeline in the $(10\bar{1}0)$ plane and the subsequent appearance in the (0002) in the AT-plots indicates twinning. It is also known that with the increase of temperature, prismatic or pyramidal slipping can also be activated. At temperatures normally higher than 150 °C dynamic recovery and recrystallization also play an important role on the deformation of magnesium. Among the possible phenomena to be observed in an AT-plot, subgrain formation also plays an important role on the deformation and is indicated by the blurring of a timeline. The spreading of the coherent domains indicates that there is an increase in misorientation, thus formation of cells and subsequently subgrains form. The diffraction results and the post experiment grain structure of the alloys shows that twinning was the predominant deformation mechanism at room temperature for all alloys in as-cast condition (Figure 6). Mg-4Nd exhibited an intensive twinned microstructure, as well as the Mg-4Nd-3Zn alloy. The twins seem thicker for the Mg-4Nd-3Zn. Twinned grains are also visible in Mg-4Nd-5Zn, but a few grains did not exhibit twins. For the Mg-4Nd-8Zn, the twins are significantly thicker compared with Mg-4Nd. The graphs of the alloys deformed at 200 °C reveal that twins are still present in most of the deformed microstructure of the alloys, although the area fraction seems to reduce substantially for Mg-4Nd-3Zn and especially Mg-4Nd-5Zn, in which twins are barely observed. Nevertheless, intensive subgrain formation indicated by the gradual change of colour within the grain is observed for the alloys compressed at 200 °C and is notable for the Mg-4Nd-5Zn alloy.

The alloys in solution treated condition (Figure 7) exhibit local twinning with regions that were subjected to higher levels of localised stresses. This indicates that the strain is not homogeneously accommodated by the microstructure. This suggests that the reinforcement of the matrix was not effective to redistribute homogeneously the stresses during deformation. Grains were partially twinned for the Mg-4Nd alloy. At 200 °C intensive subgrain formation can be observed for all the alloys, more pronounced in the case of Zn containing ones. A few twins can be also seen, but these tend to be thicker, and also fully twinned regions can be observed.

5. Conclusions

The Mg-4Nd alloy has superior mechanical properties to the ternary alloys. The change in intermetallic morphology from a continuous intermetallic to a lamella-like intermetallic is the primary cause of the reduction in mechanical properties. Only the addition of an excessive amount of Zn (8 wt.%) reaches the properties of the base alloy. The UCS is twice as large as the UTS value tested at RT in as-cast condition. The ternary alloys do not perform better after solution treatment. The deformation at RT happens mostly through twinning, at 200 °C subgrain formation occurs. There is no notable grain rotation observed in the microstructure of the as-cast alloys indicating strong grain boundary strengthening from the intermetallic. In the case of solution treated samples the microstructure undergoes twinning in a local manner. This localization can be a result of the insufficient load distribution of the intermetallic particles as the heat treatment disintegrates its network.

Author Contributions: The preparation of alloys and samples for the mechanical testing was carried out by S.G., T.S. and R.H.B. The in situ experiments were planned by D.T. and were performed by S.G., D.T., R.H.B., T.S. and A.S. all together. The post mortem characterization of the samples was done by S.G. and R.H.B. The ex situ mechanical tests were carried out by S.G. The manuscript was prepared by S.G., D.T., R.H.B., T.S. and A.S.

Funding: This research was funded by Deutsche Forschungsgemeinschaft under the grants TO817/4-1 and ME4487/1-1.

Acknowledgments: The authors acknowledge the Deutsches Elektronen-Synchrotron for the provision of facilities within the framework of the Proposal I-20150471.

Conflicts of Interest: The authors declare no conflict of interest.

References

1. Pekguleryuz, M.; Kainer, K.; Kaya, A. *Fundamentals of Magnesium Alloy Metallurgy*; Woodhead: Philadelphia, PA, USA, 2013.
2. Avedesian, M.M.; Baker, H. *Magnesium and Magnesium Alloys*; ASM Specialty Handbook; ASM International: Materials Park, OH, USA, 1999.
3. Feyerabend, F.; Fischer, J.; Holtz, J.; Witte, F.; Willumeit, R.; Drücker, H.; Vogt, C.; Hort, N. Evaluation of short-term effects of rare earth and other elements used in magnesium alloys on primary cells and cell lines. *Acta Biomater.* **2010**, *6*, 1834–1842. [CrossRef] [PubMed]
4. He, S.M.; Peng, L.M.; Zeng, X.Q.; Ding, W.J.; Zhu, Y.P. Comparison of the microstructure and mechanical properties of a ZK60 alloy with and without 1.3 wt.% gadolinium addition. *Mater. Sci. Eng. A* **2006**, *433*, 175–181. [CrossRef]
5. Li, Q.; Wang, Q.; Wang, Y.; Zeng, X.; Ding, W. Influence of temperature and strain rate on serration type transition in NZ31 Mg alloy. *J. Alloy. Compd.* **2007**, *427*, 115–123. [CrossRef]
6. Homma, T.; Mendis, C.L.; Hono, K.; Kamado, S. Effect of Zr addition on the mechanical properties of as-extruded Mg-Zn-Ca-Zr alloys. *Mater. Sci. Eng. A* **2010**, *527*, 2356–2362. [CrossRef]
7. Chang, S.Y.; Tezuka, H.; Kamio, A. Mechanical Properties and Structure of Ignition-Proof Mg-Ca-Zr Alloys Produced by Squeeze Casting. *Mater. Trans.* **1997**, *38*, 526–535. [CrossRef]
8. Ma, C.; Liu, M.; Wu, G.; Ding, W.; Zhu, Y. Tensile properties of extruded ZK60-RE alloys. *Mater. Sci. Eng. A* **2003**, *349*, 207–212. [CrossRef]
9. Zhou, H.T.; Zhang, Z.D.; Liu, C.M.; Wang, Q.W. Effect of Nd and Y on the microstructure and mechanical properties of ZK60 alloy. *Mater. Sci. Eng. A* **2007**, *445–446*, 1–6. [CrossRef]
10. Stanford, N.; Atwell, D.; Beerb, A.; Davies, C.; Barnett, M.R. Effect of microalloying with rare-earth elements on the texture of extruded magnesium-based alloys. *Scr. Mater.* **2008**, *59*, 772–775. [CrossRef]
11. Langelier, B.; Nasiri, A.M.; Lee, S.Y.; Gharghouri, M.A.; Esmaeili, S. Improving microstructure and ductility in the Mg-Zn alloy system by combinational Ce-Ca microalloying. *Mater. Sci. Eng. A* **2015**, *620*, 76–84. [CrossRef]

12. Leontis, T.E. The properties of sand cast magnesium-rare earth alloys. *J. Met.* **1949**, *185*, 968–983. [CrossRef]
13. Rokhlin, L.L. *Magnesium Alloys Containing Rare Earth Metals*; Taylor & Francis: London, UK, 2003; ISBN 0-415-28414-7.
14. Fu, P.H.; Peng, L.M.; Jiang, H.Y.; Chang, J.W.; Zhai, C.Q. Effects of heat treatments on the microstructures and mechanical properties of Mg-3Nd-0.2Zn-0.4Zr (wt.%) alloy. *Mater. Sci. Eng. A* **2008**, *486*, 183–192. [CrossRef]
15. Wu, D.; Chen, R.S.; Ke, W. Microstructure and mechanical properties of a sand-cast Mg-Nd-Zn alloy. *Mater. Des.* **2014**, *58*, 324–331. [CrossRef]
16. Zhang, J.; Li, H.; Wang, W.; Huang, H.; Pei, J.; Qu, H.; Yuan, G.; Li, Y. The degradation and transport mechanism of a Mg-Nd-Zn-Zr stent in rabbit common carotid artery: A 20-month study. *Acta Biomater.* **2018**, *69*, 372–384. [CrossRef] [PubMed]
17. Lonardelli, I.; Gey, N.; Wenk, H.R.; Humbert, M.; Vogel, S.C.; Lutterotti, L. In situ observation of texture evolution during α→β and β→α phase transformations in titanium alloys investigated by neutron diffraction. *Acta Mater.* **2007**, *55*, 5718–5727. [CrossRef]
18. Suwanpinij, P.; Stark, A.; Li, X.; Römer, F.; Herrmann, K.; Lippmann, T.; Bleck, W. In Situ High Energy X-ray Diffraction for Investigating the Phase Transformation in Hot Rolled TRIP-Aided Steels. *Adv. Eng. Mater.* **2014**, *16*, 1044–1051. [CrossRef]
19. Liss, K.D.; Yan, K. Thermo-mechanical processing in a synchrotron beam. *Mater. Sci. Eng. A* **2010**, *528*, 11–27. [CrossRef]
20. Buzolin, R.H.; Tolnai, D.; Mendis, C.L.; Stark, A.; Schell, N.; Pinto, H.; Kainer, K.U.; Hort, N. In situ synchrotron radiation diffraction study of the role of Gd, Nd on the elevated temperature compression behavior of ZK40. *Mater. Sci. Eng. A* **2015**, *640*, 129–136. [CrossRef]
21. Buzolin, R.H.; Mendis, C.L.; Tolnai, D.; Stark, A.; Schell, N.; Pinto, H.; Kainer, K.U.; Hort, N. In situ synchrotron radiation diffraction investigation of the compression behaviour at 350 °C of ZK40 alloys with addition of CaO and Y. *Mater. Sci. Eng. A* **2016**, *664*, 2–9. [CrossRef]
22. Gavras, S.; Subroto, T.; Buzolin, R.H.; Hort, N.; Tolnai, D. The Role of Zn Additions on the Microstructure and Mechanical Properties of Mg-Nd-Zn Alloys. *Int. J. Metalcast.* **2017**. [CrossRef]
23. Elsayed, F.R.; Hort, N.; Salgado Ordorica, M.A.; Kainer, K.U. Magnesium Permanent Mold Castings Optimization. *Mater. Sci. Forum* **2011**, *690*, 65–68. [CrossRef]
24. DIN 50125, *Testing of Metallic Materials—Tensile Test Pieces*; DIN: Berlin, Germany, 2009.
25. Tolnai, D.; Szakács, G.; Requena, G.; Stark, A.; Schell, N.; Kainer, K.; Hort, N. Study of the Solidification of AS Alloys Combining in situ Synchrotron Diffraction and Differential Scanning Calorimetry. *Mater. Sci. Forum* **2013**, *765*, 286–290. [CrossRef]
26. Easton, M.A.; Gibson, M.A.; Qiu, D.; Zhu, S.M.; Gröbner, J.; Schmid-Fetzer, R.; Nie, J.F.; Zhang, M.X. The role of Crystallography and Thermodynamics on Phase Selection in Binary Magnesium-Rare Earth (Ce or Nd) Alloys. *Acta Mater.* **2012**, *60*, 4420–4430. [CrossRef]
27. Zhang, B.; Gavras, S.; Nagasekhar, A.V.; Caceres, C.H.; Easton, M.A. The Strength of the Spatially Interconnected Eutectic Network in HPDC Mg-La, Mg-Nd, and Mg-La-Nd Alloys. *Metall. Mater. Trans. A* **2014**, *45*, 4386–4397. [CrossRef]

© 2018 by the authors. Licensee MDPI, Basel, Switzerland. This article is an open access article distributed under the terms and conditions of the Creative Commons Attribution (CC BY) license (http://creativecommons.org/licenses/by/4.0/).

Article

Achieving High Strength and Good Ductility in As-Extruded Mg–Gd–Y–Zn Alloys by Ce Micro-Alloying

Zhengyuan Gao [1,*], Linsheng Hu [1], Jinfeng Li [1], Zhiguo An [1], Jun Li [1] and Qiuyan Huang [2,*]

1. School of Mechatronics and Automotive Engineering, Chongqing Jiaotong University, Chongqing 400074, China; linshenghucqjtu@163.com (L.H.); ljfteemo@163.com (J.L.); anzhiguo@cqjtu.edu.cn (Z.A.); cqleejun@cqjtu.edu.cn (J.L.)
2. Institute of Metal Research, Chinese Academy of Sciences, Shenyang 110016, China
* Correspondence: zhengyuangao@cqjtu.edu.cn (Z.G.); qyhuang16b@imr.ac.cn (Q.H.)

Received: 10 December 2017; Accepted: 3 January 2018; Published: 10 January 2018

Abstract: In this study, the effect of Ce additions on microstructure evolution of Mg–7Gd–3.5Y–0.3Zn (wt %) alloys during the casting, homogenization, aging and extrusion processing are investigated, and novel mechanical properties are also obtained. The results show that Ce addition promotes the formation of long period stacking ordered (LPSO) phases in the as-cast Mg–Gd–Y–Zn–Ce alloys. A high content of Ce addition would reduce the maximum solubility of Gd and Y in the Mg matrix, which leads to the higher density of Mg12Ce phases in the as-homogenized alloys. The major second phases observed in the as-extruded alloys are micron-sized bulk LPSO phases, nano-sized stripe LPSO phases, and broken Mg12Ce and Mg5RE phases. Recrystallized grain size of the as-extruded 0.2Ce, 0.5Ce and 1.0Ce alloys can be refined to ~4.3 μm, ~1.0 μm and ~8.4 μm, respectively, which is caused by the synthesized effect of both micron phases and nano phases. The strength and ductility of as-extruded samples firstly increase and then decrease with increasing Ce content. As-extruded 0.5Ce alloy exhibits optimal mechanical properties, with ultimate strength of 365 MPa and ductility of ~15% simultaneously.

Keywords: magnesium alloys; alloying; second phases; dynamic recrystallization; mechanical properties

1. Introduction

In recent years, the demand for environment-friendly structural materials with light weight and novel mechanical properties has become urgent owing to strict emission limitations in the international community and the rapid depletion of fossil fuels [1,2]. Magnesium alloy, as the lightest metallic structural material, possesses some obvious advantages such as low density, high specific strength, notable shock absorption and noise reduction, excellent electromagnetic shielding performance, good machine ability, and recyclability [3,4]. In this sense, Mg alloy has become a promising candidate for use in various fields such as aircraft, automotives, 3D products and so on [5–7]. However, the industrial application of Mg alloy is still limited due to its low absolute strength and poor ductility [8]. Therefore, a popular research topic has been about the development of a new type of Mg alloy with high strength and excellent ductility, simultaneously.

The addition of rare earth (RE) elements plays an important role in purifying molten alloys and refining grains in Mg alloys. Moreover, it can induce formations of the high-density long period stacking ordered structure (LPSO) phases. Therefore, the properties of Mg alloy such as yield strength, ultimate strength, elongation and corrosion resistances, can be improved significantly with the addition of RE elements [9–13]. At present, Mg alloys containing the LPSO phases mainly involve ternary

Mg–Zn–X alloys (X = Y, Gd, Dy, Er, Ho, Tm, Tb), as well as the quaternary Mg–Gd–Y–Zn alloys [14,15]. It is believed that the high density of the micron-sized LPSO phases induced by multi-alloying can improve the strength of the Mg alloys via the short-fiber strengthening mechanism, while the nano-sized LPSO phases in the Mg matrix can contribute to the strength and plasticity of the alloy simultaneously based on the dislocation mechanism [16–20]. On the other hand, previous studies have shown that the volume fraction of the micron-sized and nano-sized LPSO phases increased when the other types of the micro-alloying elements are added in the Mg–Gd–Y–Zn alloy, and the strength and ductility can be further improved. For example, Wang et al. confirmed that the LPSO phases in the as-cast Mg–2.5Zn–2.5Y–1Mn (at %) alloys were increased by Ca addition [21]. Recently, Zhang et al. showed that a small amount of Zr addition could induce a new nano-sized phase in the Mg–Gd–Zn alloy, and also change the morphology and distribution of LPSO phases, while little effect of Zr addition on the quantity of the LPSO phases can be detected [21]. In particular, Ce is an alloying element that can effectively refine the Mg matrix and also greatly reduce the stacking fault energy of the α-Mg matrix [22]. In this sense, Ce may have a significant effect on formation of the LPSO phases in the RE-containing Mg alloys. However, there are few reports on the effect of Ce alloying on the microstructure and mechanical properties of Mg–Gd–Y–Zn alloy. Therefore, in this paper, the second phase in as-cast, as-homogenized, as-aged and as-extruded Mg–Gd–Y–Zn–Ce alloys is systematically investigated, and the grain size, textures and yield strength of the alloys are also characterized, in order to provide theoretical support for the development of new high-strength and high-ductility Mg wrought alloys.

2. Experiment

The as-cast Mg–7Gd–3.5Y–0.3Zn–xCe (x = 0.2, 0.5, 1.0) billets were prepared by vacuum induction melting, and this was named the 0.2Ce alloy, for example, with 0.2 wt % Ce addition. The raw materials were pure magnesium (99.95 wt %), pure zinc (99.99 wt %), pure cerium (99.97 wt %), magnesium–yttrium master alloy (Mg–20 wt % Y) and Mg–Gd master alloy (Mg–20 wt % Gd). During the melting process, argon gas was used to avoid oxidation in the furnace. The molten alloys were poured into a preheated iron mold (350 °C) when the raw materials had been melted and fully stirred. Chemical analysis was performed to determine the actual compositions, and the results are listed in Table 1.

Table 1. Compositions of the as-cast Mg–7Gd–3.5Y–0.3Zn–xCe (x = 0.2, 0.5, 1.0) alloys in weight percentage.

Samples	Gd	Y	Zn	Ce
0.2Ce alloy	7.68	3.77	0.3	0.21
0.5Ce alloy	6.77	3.42	0.3	0.47
1.0Ce alloy	7.55	3.69	0.28	1.05

Homogenization treatments were conducted on the as-cast ingots at the temperature of 540 °C for 24 h. As-aged samples were prepared by a subsequent aging treatment at 400 °C for 1 h. After pre-heating at 400 °C for 20 min, the as-aged samples were indirectly extruded into the bars with a diameter of 10 mm under an extrusion ratio of 18:1, extrusion temperature of 400 °C and a ram speed of 0.5 mm/s. Fiel-emission scanning-electron microscopy (SEM) and an energy-dispersive spectrometer (EDS, JEOL JEM-2100F, JEOL Ltd., Osaka, Japan) were applied to reveal the microstructure of as-cast, as-homogenized, as-aged and as-extruded samples, and also the compositions of the second phases. The microstructures of the as-extruded samples were also characterized by optical microscopy (OM, ZEISS, Heidenheim, Germany), and the macro-texture of the as-extruded samples were determined by X-ray diffraction (XRD, Philips PW3040/60 X' Pert PRO, Royal Dutch Philips Electronics Ltd., Amsterdam, The Netherlands). Finally, the mechanical properties of the samples (length of 25 mm, diameter of 5 mm) were tested under tension directions, with the initial strain rate of 10−3/s

(Shimazu AG-X Plus, SHIMAZU, Kyōto-fu, Japan). The equilibrium phase diagrams of the Mg alloys were also calculated using the Panda software (Compu Therm LLC, Madison, WI, USA).

3. Results

3.1. Microstructures of the As-Cast Mg–Gd–Y–Zn–Ce Alloys

The SEM images of the as-cast Mg–7Gd–3.5Y–0.3Zn–xCe (x = 0.2, 0.5, 1.0, wt %) alloys are shown in Figure 1. The as-cast 0.2Ce, 0.5Ce and 1.0Ce alloys exhibit a typical non-equilibrium solidification microstructure, containing the dendrite and dendritic segregation zones. By increasing Ce additions, the number density of irregular-shaped phases and stripe-phases increases and the size also grows gradually. As listed in Table 2, the EDS results indicate that the composition of irregular 1# second phase in the 0.2Ce as-cast alloy is Mg89.38Gd5.36Y3.5Zn1.12Ce0.63, and the 2# stripe second phase contains 2.98 at % RE (Gd, Y, Ce) and 2.54 at % Zn, which are indicated by circles in Figure 1b. Based on the morphologies and atomic ratios (RE/Zn) of the compounds, the corresponding 1# and 2# particles can be identified to be the Ce-enriched Mg_5RE phase and the LPSO phase (RE/Zn ≈ 1), respectively. These results are in agreement with previous reports on the second phases formed in the as-cast Mg–Gd–Y–Zn based alloys [12,13]. Moreover, EDS analysis of the α-Mg matrix shows higher content of dissolved Ce solute with increasing Ce additions.

Table 2. Energy-dispersive spectrometry (EDS) results in as-cast Mg–7Gd–3.5Y–0.3Zn–xCe (x = 0.2, 0.5, 1.0) alloys (at %).

Points	Mg	Gd	Y	Zn	Ce
1#	89.38	5.36	3.50	1.12	0.63
2#	94.47	1.11	1.86	2.54	0.01
3#	88.29	5.94	4.65	0.55	0.58
4#	93.0	1.46	1.72	3.61	0.22
5#	87.14	5.46	4.19	1.74	1.47
6#	92.01	1.89	1.82	3.90	0.38

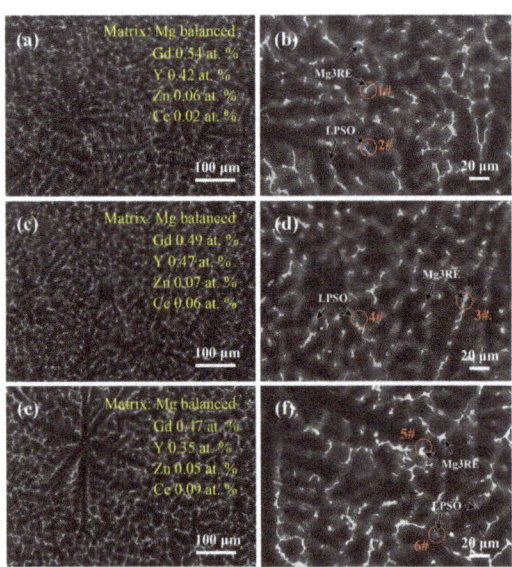

Figure 1. Scanning-electron microscopy (SEM) images of the as-cast Mg–7Gd–3.5Y–0.3Zn–xCe (x = 0.2, 0.5, 1.0) alloys: (a,b) 0.2Ce alloy; (c,d) 0.5Ce alloy; (e,f) 1.0Ce alloy.

The size and distribution of the Ce-enriched Mg$_5$RE phase and LPSO phase in the as-cast 0.5Ce alloy is similar to that in the as-cast 0.2Ce alloy, while the fraction of the LPSO phase and the dissolved content of Ce atoms increase slightly. In the as-cast 1Ce alloy, on the other hand, the size of the Ce-enriched Mg$_5$RE phase increases significantly, and the quantity of the LPSO phase also becomes obviously higher.

3.2. Microstructures of the As-Homogenized and As-Aged Mg–Gd–Y–Zn–Ce Alloys

Figure 2 shows the SEM and optical microscopy images of the as-homogenized 0.2Ce, 0.5Ce and 1.0Ce alloys. The second phases and the segregation zones in the as-cast 0.2Ce and 0.5Ce alloys have been almost dissolved by the high temperature solid solution treatment. In contrast, a large number of reticular and spherical second phases still exist along the grain boundaries and within the grain interior of the as-homogenized 1.0Ce alloy. According to the EDS results, the second phase is the Mg$_{12}$Ce phase enriched with RE elements. At the same time, the grain size of as-homogenized 0.2Ce, 0.5Ce and 1.0Ce samples decreases gradually with the increasing addition of Ce, i.e., ~370 μm, ~260 μm and ~160 μm, respectively. As shown in Figure 3, after the subsequent aging treatment of the as-homogenized samples, numerous precipitations can be detected in the 0.2Ce, 0.5Ce and 1.0Ce alloys, including the reticular, spherical and stripe phases. The EDS results (Table 3) show that the spherical second phase "a" in Figure 3b contains ~3.4 at % Gd, ~9.21 at % Y and a small amount of Zn and Ce; the stripe second phase "b" contains ~2.73 at % Gd/Y, ~2.28 at % Zn and a small amount of Ce (RE/Zn ≈ 1); the reticular second phase "c" contains ~2.7 at % Gd/Y, ~1.5 at % Ce and a small amount of Zn. Therefore, the second phase "a" and "c" that precipitates in the as-aged 0.2Ce alloy can be clarified to be the Ce-enriched Mg$_5$RE phase, while the second phase "b" is determined as the LPSO phase. The length and volume fraction of nano-LPSO phase in as-aged 0.2Ce alloy is estimated to be ~7 μm and ~2%, respectively.

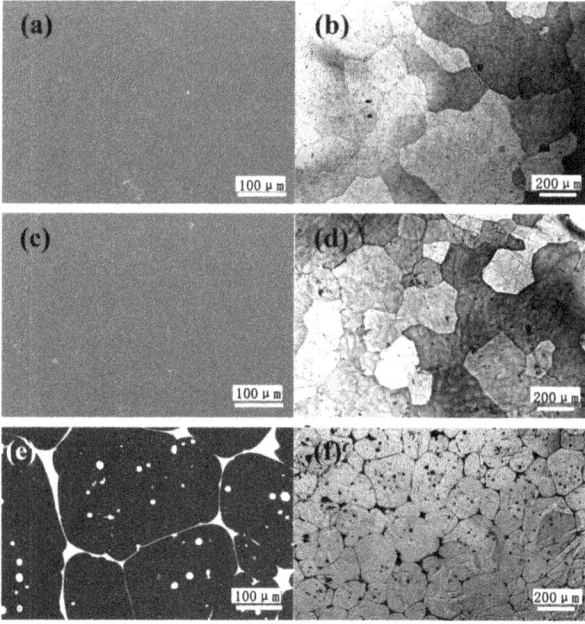

Figure 2. (a,c,e) SEM images and (b,d,f) optical images of the as-homogenized Mg–7Gd–3.5Y–0.3Zn–xCe (x = 0.2, 0.5, 1.0) alloys: (a,b) 0.2Ce alloy; (c,d) 0.5Ce alloy; (e,f) 1.0Ce alloy.

Table 3. EDS results in as-aged Mg–7Gd–3.5Y–0.3Zn–xCe (x = 0.2, 0.5, 1.0) alloys (at %).

Points	Mg	Gd	Y	Zn	Ce
a	86.79	3.4	9.21	0.46	0.15
b	94.94	1.04	1.69	2.28	0.05
c	97.13	1.20	1.50	0.09	1.5
d	87.00	3.05	9.34	0.23	0.38
e	92.62	1.1	2.28	3.38	0.62
f	86.57	4.47	3.80	3.27	3.89
g	80.46	7.07	6.64	0.5	5.33
h	58.87	8.94	30.45	0.66	1.08
i	88.13	3.64	2.94	3.06	2.24

With increasing content of Ce, both the size and number density of the second phases precipitated from the 0.5Ce alloy and the 1.0Ce alloy increase significantly (Figure 3e,h). The EDS results (Table 3) indicate that the spherical phase "d" in 0.5Ce alloy should be the Ce-enriched Mg_5RE phases due to the high RE content, and the stripe second phases "e" are also the LPSO phases, while the reticular phase "f" should be the $Mg_{12}Ce$ phase due to the high Ce content. In other words, the addition of Ce would induce formation of the new phase of $Mg_{12}Ce$ in present Mg–Gd–Y–Zn alloys. The quantity of the $Mg_{12}Ce$ phases and the LPSO phases in 1.0Ce alloy increases significantly, as shown in Figure 3h. Volume fractions of the nano-LPSO phases in as-aged 0.5Ce and 1.0Ce alloys have increased to ~4% and ~9%, respectively. Moreover, the EDS analysis shows that the remaining Gd and Y elements dissolved in the α-Mg matrix obviously decrease with increasing Ce additions.

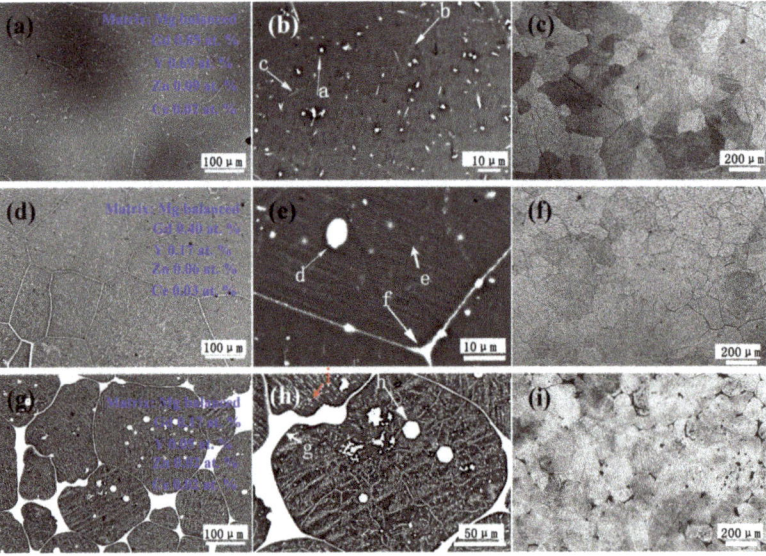

Figure 3. (a,b,d,e,g,h) SEM images and (c,f,i) optical images of the as-aged Mg–7Gd–3.5Y–0.3Zn–xCe (x = 0.2, 0.5, 1.0) alloys: (a,b,c) 0.2Ce alloy; (d,e,f) 0.5Ce alloy; (g,h,i) 1.0Ce alloy; (a,d,g) lower-magnification images; and (b,e,h) higher-magnification images.

3.3. Microstructures and Mechanical Properties of the As-Extruded Mg–Gd–Y–Zn–Ce Alloys

Figure 4 shows the microstructures of as-extruded 0.2Ce, 0.5Ce and 1.0Ce alloys. As compared with the as-homogenized samples, a large number of micron-size second phases precipitate in the as-extruded alloys and distribute along the extrusion direction after the hot extrusion process (Figure 4a,d,g). Optical images show that dynamic recrystallization (DRX) occurs in the as-extruded 0.2Ce, 0.5Ce and 1.0Ce alloys, and volume fraction of the recrystallized grains is ~65%, ~75% and ~97%, respectively (Figure 4b,e,h); that is, the proportion of recrystallization increases with the increasing addition of Ce. At the same time, the coarsely as-deformed regions are compressed to be the streamline, which is aligned with the extrusion direction. The DRXed grain sizes of the 0.2Ce, 0.5Ce and 1.0Ce alloys are estimated to be ~4.3 μm, ~1.0 μm and ~8.4 μm, respectively, according to the higher magnification optical images (Figure 4c,f,i). It can be noted that grain sizes of the non-DRXed grains are not considered due to the severely heterogeneous microstructure. On the other hand, there usually exists a relationship between the macro-texture and the dynamic recrystallization behavior of the as-extruded Mg alloys. In general, DRXed grains of the RE-containing Mg alloys display a "rare earth" texture, in which the c-axis deviates from the radial direction of the extrusion rod to the extrusion direction [15,23]. Therefore, the more dynamic the recrystallization that takes place in rare earth Mg alloys, the more obvious the "rare earth" texture that would be generated. As shown in Figure 5, the macro-texture of the as-extruded 0.2Ce, 0.5Ce and 1.0Ce alloys agrees well with the statement above. Due to the high proportion of recrystallized grains and the concurrent existence of non-recrystallized grains, the "rare earth" texture and also the fiber texture co-exist in the as-extruded 0.2Ce alloy (Figure 5a). The fiber texture shows the typical texture of the non-recrystallized grains in which the c-axis is aligned with the radial direction of the bar [23].

With increasing addition of Ce, a higher degree of recrystallization takes place, and the c-axis tilts to the extrusion direction gradually, as shown in Figure 5b,c. The fiber texture component in the 1.0Ce alloy has basically disappeared, and only the "rare earth" texture remains, which is consistent with the complete recrystallization as detected in Figure 4.

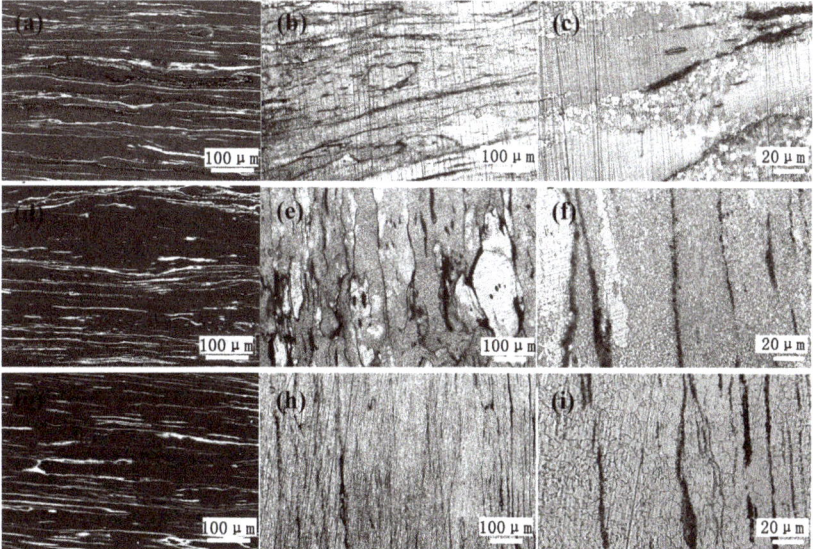

Figure 4. (a,d,g) SEM images and (b,c,e,f,h,i) optical images of the as-extruded Mg–7Gd–3.5Y–0.3Zn–xCe (x = 0.2, 0.5, 1.0) alloys: (a,b,c) 0.2Ce alloy; (d,e,f) 0.5Ce alloy; (g,h,i) 1.0Ce alloy; (b,e,h) lower-magnification images; and (c,f,i) higher-magnification images.

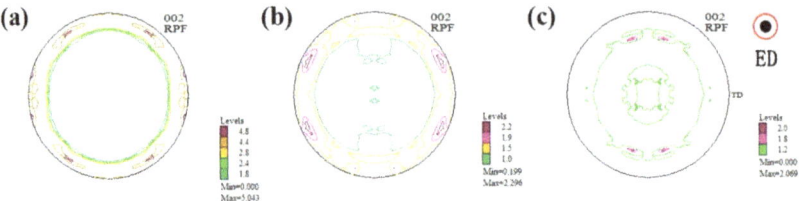

Figure 5. The (0001) pole figures of the as-extruded Mg–7Gd–3.5Y–0.3Zn–xCe (x = 0.2, 0.5, 1.0) alloys: (a) 0.2Ce alloy; (b) 0.5Ce alloy; (c) 1.0Ce alloy.

As mentioned above, the morphology of the second phases varies with different amounts of Ce addition in the as-homogenized samples. Accordingly, different precipitation behaviors of the second phases are also generated in the as-extruded alloys, as shown in SEM images of as-extruded 0.2Ce, 0.5Ce and 1.0Ce alloys (Figure 6). Cross-sectional scanning results show that a large number of the non-fragmented reticular and spherical micron-size second phases are uniformly distributed in the matrix (Figure 6a,c,e), which coincides with the results of longitudinal-section scanning in Figure 4. Based on the characteristic "twisted" morphology of the second phases, these bulk micron-size phases can be identified as LPSO phases [12,13]. In order to clarify the effect of the second phase on dynamic recrystallization, higher-magnification images are also displayed and a large number of nano-sized stripe second phase can be found to disperse in the grain interior of the Mg alloys (Figure 6b,d,f). The nano-phases are similar to the nano-LPSO phases formed in the as-aged samples. Thus, it can be concluded that during hot extrusion deformation, both bulk LPSO phases and stripe nano-LPSO phases precipitated from the matrices of the 0.2Ce, 0.5Ce and 1.0Ce alloys. However, both the size and distribution of the bulk and nano-LPSO phases in the three groups of alloys are different from each other. The density and fraction of the stripe LPSO phases are high in the 0.2Ce and 0.5Ce alloys, while the stripe LPSO phase exists in the form of clusters with lower inter-distances in the 1.0Ce alloy and the volume fraction obviously decreases (Table 4). Moreover, the micron-sized $Mg_{12}Ce$ phases are broken into spherical particles in the 0.5Ce and 1.0Ce alloys, and co-exist with the Mg_5RE phases. With increasing Ce addition, the corresponding volume fraction of the spherical phases also increases.

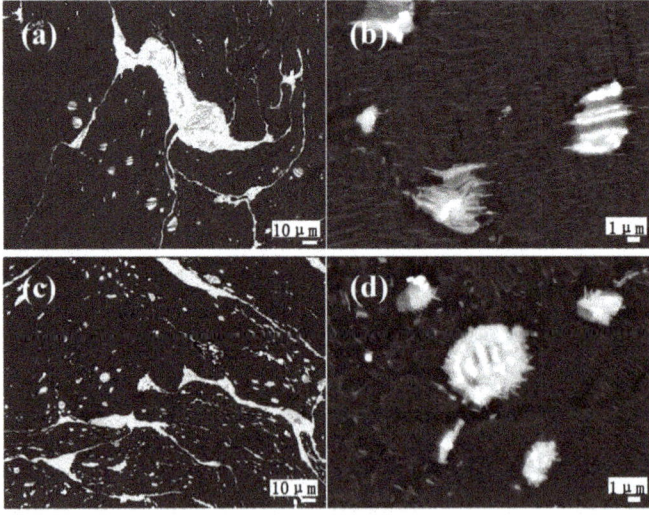

Figure 6. Cont.

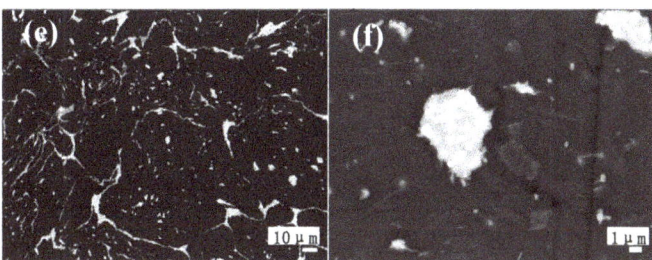

Figure 6. SEM images of the as-extruded Mg–7Gd–3.5Y–0.3Zn–xCe (x = 0.2, 0.5, 1.0) alloys: (**a,b**) 0.2Ce alloy; (**c,d**) 0.5Ce alloy; (**e,f**) 1.0Ce alloy.

Table 4. Sizes of nano-long period stacking ordered structure (LPSO) and bulk LPSO, the fraction of second phases and the DRXed grain size of as-extruded Mg–7Gd–3.5Y–0.3Zn–xCe (x = 0.2, 0.5, 1.0) alloys.

As-Extruded Samples	Length of Nano-LPSO Phase	Length of Bulk LPSO Phase	Width of Bulk LPSO Phase	Fraction of Second Phases	DRXed Grain Sizes
0.2Ce alloy	~3.0 μm	~300 μm	~35 μm	4% (nano) 10% (micron)	~4.3 μm
0.5Ce alloy	~1.5 μm	~260 μm	~25 μm	5% (nano) 11% (micron)	~1.7 μm
1.0Ce alloy	~1.0 μm	~200 μm	~15 μm	2% (nano) 16% (micron)	~8.4 μm

The mechanical properties of the as-extruded samples under both tensile and compressive conditions are shown in Figure 7 and Table 5. Under tensile deformation, the yield strength (YS) and ultimate strength (UTS) of as-extruded 0.2Ce alloy are 286.4 MPa and 340.7 MPa, respectively. With the addition of 0.5 wt % Ce, the YS and UTS increase to 302.2 MPa and 365.3 MPa. For as-extruded 1.0Ce alloy, the YS and UTS decrease to 263.5MPa and 327.3MPa, respectively. In addition, the elongation to fracture of the as-extruded alloy increases from ~13% in 0.2Ce alloy to ~15% in 0.5Ce alloy, and then decreases to ~12% in the as-extruded 1.0Ce alloy. The strength and ductility of the as-extruded alloys exhibit a trend of initial increase and then decrease with increasing Ce addition. The YS, UTS and elongation of the as-extruded 0.5Ce alloy are the highest among the three samples. In other words, the addition of a small amount of Ce is beneficial for improving mechanical properties of as-extruded Mg–7Gd–3.5Y–0.3Zn alloys, and the optimal Ce content is about 0.5 wt %.

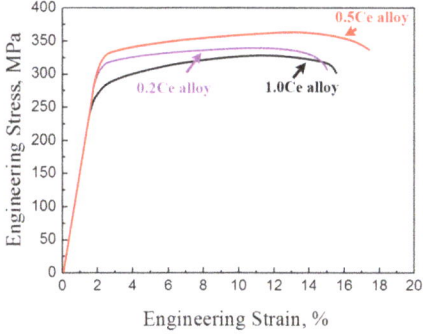

Figure 7. Engineering stress-strain curves of the as-extruded Mg–7Gd–3.5Y–0.3Zn–xCe (x = 0.2, 0.5, 1.0) alloys.

Table 5. Mechanical properties of as-extruded Mg–Gd–Y–Zn–Ce alloys under tension.

Samples	Yield Strength (MPa)	Ultimate Strength (MPa)	Elongation (%)
0.2Ce alloy (tension)	286.4	340.7	13
0.5Ce alloy (tension)	302.2	365.3	15
1.0Ce alloy (tension)	263.5	327.3	13.5

4. Discussion

As-cast samples exhibit the typical non-equilibrium microstructure, and the main second phases are Ce-enriched Mg_5RE and LPSO phases. With the addition of Ce, not only does the size of Mg_5RE phases increase, but also the fraction of LPSO phases increases gradually, which means that Ce addition can promote the formation of LPSO phases in the as-cast Mg–Gd–Y–Zn alloys. This may be attributed to the mechanism whereby Ce reduces the stacking fault energy of the Mg matrix effectively [22]. After solid solution treatment, the eutectic microstructures of as-cast 0.2Ce and 0.5Ce alloys have almost disappeared, while large numbers of RE-enriched $Mg_{12}Ce$ phases remain in the 1.0Ce alloy. Consequently, the maximum solid solubility of Gd and Y in the Mg matrix would be reduced because of the addition of a large amount of Ce, as confirmed by the EDS analysis on the α-Mg matrix. In fact, Figure 8 displays the polythermic sections for the equilibrium phase diagrams of the Mg–0.3Zn–0.2Ce–xGd (x = 0–10 wt %) and Mg–0.3Zn–1.0Ce–xGd (x = 0–10 wt %) alloys. It can clearly be seen that Ce addition would decrease the maximum solubility of the Gd elements, and also induce formation of the new $Mg_{12}Ce$ phase.

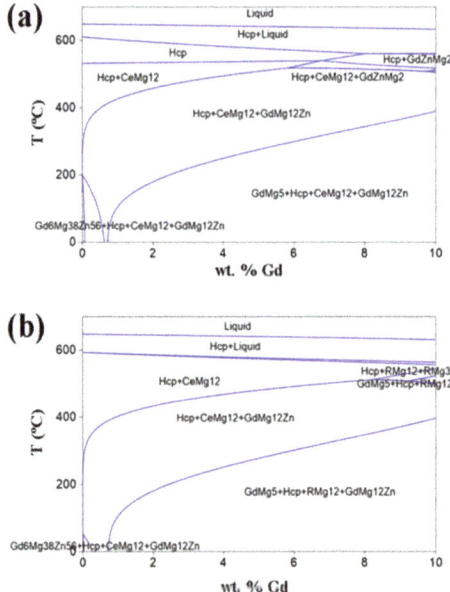

Figure 8. The polythermic sections for the equilibrium phase diagrams of the (a) Mg–0.3Zn–0.2Ce–xGd (x = 0–10 wt %) and (b) Mg–0.3Zn–1.0Ce–xGd (x = 0–10 wt %) alloys.

During the subsequent extrusion process, the bulk micron-sized LPSO phases dynamically precipitate at the grain interiors and boundaries of the as-received 0.2Ce, 0.5Ce and 1.0Ce alloys, as shown in Figure 6. The micron-sized $Mg_{12}Ce$ and Mg_5RE phases in the 0.5Ce and 1.0Ce alloys have been broken into spherical particles in the conditions of thermo-mechanical processing. Consistent with the mechanism of particle-stimulated nucleation (PSN), these micron-sized particles will promote the dynamic recrystallization of the Mg matrix [15,23]. On the other hand, a large number of nano-sized stripe LPSO phases also precipitate at the grain interiors (Table 4). The second phases with inter-spacing of ~100 nm can bring an obvious drag effect on the DRXed grain boundaries, and then growths of the DRXed grains can be inhibited [12,13]. After the extrusion deformation, the 0.2Ce alloy exhibits a bimodal grain structure composed of coarse grains and ultra-fine grains. When the content of Ce is increased to 0.5 wt %, the volume fraction of the bulk LPSO phase, the micron-sized $Mg_{12}Ce$ and the Mg_5RE phases increase. As a result, the nucleation sites of dynamic recrystallization increase. At the same time, a high content of Ce promotes the dynamic precipitation of the higher density of stripe LPSO phases (Table 4), and enhances the dragging effect on the mobility of the grain boundaries. Consequently, a higher fraction of recrystallization occurs, and the ultrafine grains (only ~1.0 μm) are formed in the as-extruded 0.5Ce alloy. With the addition of Ce increasing to 1 wt %, more Gd, Y and other elements will combine with Ce, in the form of bulk LPSO and Mg_5RE phases, which exist in the interior or boundaries of the grains. As a result, the driving pressure for recrystallization via the PSN mechanism increases during the extrusion process. However, the precipitation of a larger number of bulk LPSO phases leads to the decrease of the residual Gd and Y elements in the Mg matrix (Table 4), and consequently the fraction of nano-sized stripe LPSO phase is obviously reduced. The inhibition effect on growth of the recrystallized grains is weakened, which leads to the formation of larger recrystallized grains in 1.0Ce alloy (~8.5 μm).

Grain refinement is an effective approach for improving the strength and ductility of Mg alloys [24]. The recrystallized grains of the as-extruded 0.2Ce alloy can be refined to ~4.3 μm, and thus they can exhibit high strength and ductility simultaneously. At the same time, the as-extruded 0.2Ce alloy also displays the characteristics of a bimodal grain structure, in which the coarse grains will improve the strength and plasticity of the alloy via the back-stress hardening mechanism [25]. For instance, Wu et al. [25] revealed that, based on the bimodal grain structure, the yield strength of titanium alloy was increased by three times, and elongation was maintained at ~8%. On the other hand, there are a large number of reticular LPSO phases and nano-sized LPSO phases in the as-extruded alloys (Table 4), which would enhance the strength and ductility of the alloys by fiber-strengthening and dispersion-strengthening mechanisms [3,17]. Recently, Hagihara et al. [14] also confirmed the strengthening effect of fiber-shaped LPSO phases in the Mg alloy. When the content of Ce increases to 0.5 wt %, the size of recrystallized grains of the alloy is refined to ~1.0 μm, the yield strength of the alloy is improved effectively based on the mechanism of grain-refinement strengthening, and the plasticity can also be improved due to the existence of the bimodal grain structure. At the same time, the density of the bulk and stripe LPSO phases increase, which can further improve the strength of the alloy. With increasing Ce addition to 1 wt %, strength and ductility will be reduced due to the growth of recrystallized grains (~8.4 μm) and the lower density of the nano-sized stripe LPSO phases (~2%, Table 4). In this sense, Ce addition can change the size and distribution of second phases in present alloys, and has an effect on the dynamic recrystallization behavior during the extrusion process to then enhance the mechanical properties of the as-extruded alloys.

5. Conclusions

In this work, the effects of Ce alloying on the microstructure and mechanical properties of as-cast, as-homogenized, as-aged and as-extruded Mg–7Gd–3.5Y–0.3Zn alloys have been investigated, and the following conclusions can be drawn:

(1) Ce addition promotes the formation of LPSO phases in the as-cast Mg–Gd–Y–Zn–Ce alloys. The high content of Ce addition reduces the maximum solubility of Gd and Y in the Mg matrix, and leads to a higher density of $Mg_{12}Ce$ phases in the as-homogenized alloys;

(2) the main second phases in the as-extruded alloys are bulk micron-sized LPSO phases, nano-sized stripe LPSO phases and broken $Mg_{12}Ce$ and Mg_5RE phases. The recrystallized grain size of the as-extruded 0.2Ce, 0.5Ce and 1.0Ce alloys can be refined to ~4.3 μm, ~1.0 μm and ~8.4 μm,

respectively. This can be attributed to the synthesized effect of the driving force of the micron phases and the dragging effect of the nano phases on dynamic recrystallization;
(3) with the increasing content of Ce, tension, strength and ductility of as-extruded samples would firstly increase and then decrease. The as-extruded 0.5Ce alloy exhibits optimal mechanical properties.

Acknowledgments: This work was supported by the Project of National Nature Science Foundation of China (51401178, 51701211); the Scientific and Technological Research Program of Nanchuan in Chongqing (No. CX201407); the Scientific and Technological Research Program of Chongqing Jiaotong University (16JDKJC-A005); and the Chongqing Research Program of Basic Research and Frontier Technology (CSTC2015JCYJBX0140).

Author Contributions: Zhengyuan Gao conceived, designed and performed the experiments; Linsheng Hu, Jinfeng Li, Zhiguo An and Jun Li collected, analyzed and interpreted data; Qiuyan Huang wrote the paper and analyzed data.

Conflicts of Interest: The authors declare no conflict of interest.

References

1. You, S.; Huang, Y.; Kainer, K.U.; Hort, N. Recent research and developments on wrought magnesium alloys. *J. Magnes. Alloys* **2017**, *5*, 239–253. [CrossRef]
2. Liu, X.; Liu, Y.; Jin, B.; Lu, Y.; Lu, J. Microstructure evolution and mechanical properties of a smated Mg alloy under in situ sem tensile testing. *J. Mater. Sci. Technol.* **2017**, *33*, 224–230. [CrossRef]
3. Pan, H.; Fu, H.; Ren, Y.; Huang, Q.; Gao, Z.; She, J.; Qin, G.; Yang, Q.; Song, B.; Pan, F. Effect of Cu/Zn on microstructure and mechanical properties of extruded Mg–Sn alloys. *Mater. Sci. Technol.* **2016**, *32*, 1240–1248. [CrossRef]
4. Pan, H.; Fu, H.; Song, B.; Ren, Y.; Zhao, C.; Qin, G. Formation of profuse dislocations in deformed calcium-containing magnesium alloys. *Philos. Mag. Lett.* **2016**, *96*, 249–255. [CrossRef]
5. Pan, H.; Ren, Y.; Fu, H.; Zhao, H.; Wang, L.; Meng, X.; Qin, G. Recent developments in rare-earth free wrought magnesium alloys having high strength: A review. *J. Alloys Compd.* **2016**, *663*, 321–331. [CrossRef]
6. Pan, H.; Qin, G.; Xu, M.; Fu, H.; Ren, Y.; Pan, F.; Gao, Z.; Zhao, C.; Yang, Q.; She, J.; et al. Enhancing mechanical properties of Mg–Sn alloys by combining addition of Ca and Zn. *Mater. Des.* **2015**, *83*, 736–744. [CrossRef]
7. Huang, Q.; Tang, A.; Ma, S.; Pan, H.; Song, B.; Gao, Z.; Rashad, M.; Pan, F. Enhancing thermal conductivity of Mg–Sn alloy sheet by cold rolling and aging. *J. Mater. Eng. Perform.* **2016**, *25*, 2356–2363. [CrossRef]
8. Barnett, M.R.; Nave, M.D.; Bettles, C.J. Deformation microstructures and textures of some cold rolled mg alloys. *Mater. Sci. Eng. A* **2004**, *386*, 205–211. [CrossRef]
9. Kiani, M.; Gandikota, I.; Rais-Rohani, M.; Motoyama, K. Design of lightweight magnesium car body structure under crash and vibration constraints. *J. Magnes. Alloys* **2014**, *2*, 99–108. [CrossRef]
10. Li, T.; Zhang, K.; Li, X.; Du, Z.; Li, Y.; Ma, M.; Shi, G. Dynamic precipitation during multi-axial forging of an Mg–7Gd–5Y–1Nd–0.5Zr alloy. *J. Magnes. Alloys* **2013**, *1*, 47–53. [CrossRef]
11. Lv, B.; Peng, J.; Peng, Y.; Tang, A. The effect of addition of Nd and Ce on the microstructure and mechanical properties of ZM21 Mg alloy. *J. Magnes. Alloys* **2013**, *1*, 94–100. [CrossRef]
12. Yu, Z.; Huang, Y.; Gan, W.; Mendis, C.L.; Zhong, Z.; Brokmeier, H.G.; Hort, N.; Meng, J. Microstructure evolution of Mg–11Gd–4.5Y–1Nd–1.5Zn–0.5Zr (wt %) alloy during deformation and its effect on strengthening. *Mater. Sci. Eng. A* **2016**, *657*, 259–268. [CrossRef]
13. Yu, Z.; Huang, Y.; Mendis, C.L.; Hort, N.; Meng, J. Microstructural evolution and mechanical properties of Mg–11Gd–4.5Y–1Nd–1.5Zn–0.5Zr alloy prepared via pre-ageing and hot extrusion. *Mater. Sci. Eng. A* **2015**, *624*, 23–31. [CrossRef]
14. Hagihara, K.; Kinoshita, A.; Sugino, Y.; Yamasaki, M.; Kawamura, Y.; Yasuda, H.; Umakoshi, Y. Effect of long-period stacking ordered phase on mechanical properties of Mg97Zn1Y2 extruded alloy. *Acta Mater.* **2010**, *58*, 6282–6293. [CrossRef]
15. Yamasaki, M.; Hashimoto, K.; Hagihara, K.; Kawamura, Y. Effect of multimodal microstructure evolution on mechanical properties of Mg–Zn–Y extruded alloy. *Acta Mater.* **2011**, *59*, 3646–3658. [CrossRef]

16. Tane, M.; Nagai, Y.; Kimizuka, H.; Hagihara, K.; Kawamura, Y. Elastic properties of an Mg–Zn–Y alloy single crystal with a long-period stacking-ordered structure. *Acta Mater.* **2013**, *61*, 6338–6351. [CrossRef]
17. Hagihara, K.; Yokotani, N.; Umakoshi, Y. Plastic deformation behavior of $Mg_{12}YZn$ with 18R long-period stacking ordered structure. *Intermetallics* **2010**, *18*, 267–276. [CrossRef]
18. Garces, G.; Morris, D.G.; Muñoz-Morris, M.; Pérez, P.; Tolnai, D.; Mendis, C.; Stark, A.; Lim, H.; Kim, S.; Shell, N. Plasticity analysis by synchrotron radiation in a $Mg_{97}Y_2Zn_1$ alloy with bimodal grain structure and containing lpso phase. *Acta Mater.* **2015**, *94*, 78–86. [CrossRef]
19. Huang, Q.; Pan, H.; Tang, A.; Ren, Y.; Song, B.; Qin, G.; Zhang, M.; Pan, F. On the dynamic mechanical property and deformation mechanism of as-extruded Mg–Sn–Ca alloys under tension. *Mater. Sci. Eng. A* **2016**, *664*, 43–48. [CrossRef]
20. Pan, H.; Qin, G.; Huang, Y.; Yang, Q.; Ren, Y.; Song, B.; Chai, L.; Zhao, Z. Activating profuse pyramidal slips in magnesium alloys via raising strain rate to dynamic level. *J Alloys Compd.* **2016**, *688*, 149–152. [CrossRef]
21. Wang, J.; Zhang, J.; Zong, X.; Xu, C.; You, Z.; Nie, K. Effects of ca on the formation of lpso phase and mechanical properties of Mg–Zn–Y–Mn alloy. *Mater. Sci. Eng. A* **2015**, *648*, 37–40. [CrossRef]
22. Wang, W.Y.; Shang, S.L.; Wang, Y.; Mei, Z.G.; Darling, K.A.; Kecskes, L.J.; Mathaudhu, S.N.; Hui, X.D.; Liu, Z.K. Effects of alloying elements on stacking fault energies and electronic structures of binary Mg alloys: A first-principles study. *Mater. Res. Lett.* **2014**, *2*, 29–36. [CrossRef]
23. Oh-Ishi, K.; Mendis, C.; Homma, T.; Kamado, S.; Ohkubo, T.; Hono, K. Bimodally grained microstructure development during hot extrusion of Mg–2.4Zn–0.1Ag–0.1Ca–0.16Zr (at %) alloys. *Acta Mater.* **2009**, *57*, 5593–5604. [CrossRef]
24. Yuan, W.; Mishra, R.S. Grain size and texture effects on deformation behavior of AZ31 magnesium alloy. *Mater. Sci. Eng. A* **2012**, *558*, 716–724. [CrossRef]
25. Wu, X.; Yang, M.; Yuan, F.; Wu, G.; Wei, Y.; Huang, X.; Zhu, Y. Heterogeneous lamella structure unites ultrafine-grain strength with coarse-grain ductility. *Proc. Natl. Acad. Sci. USA* **2015**, *112*, 14501–14505. [CrossRef] [PubMed]

© 2018 by the authors. Licensee MDPI, Basel, Switzerland. This article is an open access article distributed under the terms and conditions of the Creative Commons Attribution (CC BY) license (http://creativecommons.org/licenses/by/4.0/).

Article

Maintaining High Strength in Mg-LPSO Alloys with Low Yttrium Content Using Severe Plastic Deformation

Gerardo Garces [1,*], Sandra Cabeza [2], Rafael Barea [3], Pablo Pérez [1] and Paloma Adeva [1]

1. Departamento de Metalurgia Física, Centro Nacional de Investigaciones Metalúrgicas, CENIM, CSIC, Avda. Gregorio del Amo 8, 28040 Madrid, Spain; zubiaur@cenim.csic.es (P.P.); adeva@cenim.csic.es (P.A.)
2. Institute Laue-Langevin, ILL, 38042 Grenoble, France; cabeza@ill.fr
3. Departamento de Ingeniería Industrial, Universidad Nebrija, Campus Dehesa de la Villa, C. Pirineos 55, 28040 Madrid, Spain; rbarea@nebrija.es
* Correspondence: ggarces@cenim.csic.es; Tel.: +34-91-5538900 (ext. 336)

Received: 5 April 2018; Accepted: 2 May 2018; Published: 5 May 2018

Abstract: Alternative processing routes such as powder metallurgy, the extrusion of recycled chips, or equal channel angular pressing (ECAP) have been considered for effective methods of maintaining the high mechanical strength of Mg-Y-Zn alloys containing long-period stacking ordered structures with respect to the alloy processed by the conventional extrusion of as-cast ingots with the advantage of minimizing the yttrium content. A yield stress similar to that found for extruded $Mg_{97}Y_2Zn_1$ alloy can be attained with only half of the usual yttrium and zinc additions thanks to the grain refinement induced by ECAP processing. The properties of $Mg_{98.5}Y_1Zn_{0.5}$ subjected to ECAP are maintained up to 200 °C, but superplastic behavior is found above this temperature when the alloy is processed through a powder metallurgy route.

Keywords: magnesium alloys; ECAP; powder metallurgy; mechanical properties; LPSO-phase

1. Introduction

The mechanical strength and creep resistance of Mg-Y,RE-Zn (RE = rare earths) alloys containing long-period stacking ordered (LPSO) phases at intermediate temperatures exceed by far those of commercial magnesium alloys [1–6]. LPSO-phases are solid solutions of yttrium or certain rare earth elements and transition metals in the magnesium lattice, where these atoms are arranged periodically in the magnesium basal planes to form ordered structures [7–11]. Depending on the processing route, important variations in mechanical properties can be obtained. The $Mg_{97}Y_2Zn_1$ (at %) alloy produced by the warm extrusion of rapidly solidified (RS) powders exhibited the highest high yield strength reported in this system, about 610 MPa [1]. The extrusion of RS ribbons with a very fine dendritic microstructure renders a yield stress of 410 MPa [12]. The optimization of extrusion parameters of as-cast $Mg_{97}Y_2Zn_1$ ingots results in a material combining cast high-yield stress (around 350 MPa) and appreciable ductility [3]. Furthermore, subsequent equal channel angular processing (ECAP) after extrusion leads to yield stresses higher than 400 MPa due to grain refinement of the magnesium phase [13,14]. However, in all cases, the yttrium and/or RE contents were at least 2 at %. Reduction in the use of rare earth elements is a critical point in the design of new high-strength magnesium alloys because there is a tendency to minimize the content of rare earth elements since the market is controlled by China and future availability is not guaranteed. Consequently, Europe, Japan, and the USA are proposing policies/guidelines addressed to substitute the use of such elements as much as possible [15,16]. In the case of LPSO-phase containing alloys, the decrease in the Y concentration implies a significant decrease in mechanical strength because the volume fraction of the reinforcing

LPSO-phase is reduced [17]. Therefore, an additional refinement of the grain size in alloys containing the LPSO-phase is needed in such a way that it can compensate minor hardening due to the smaller volume fraction of the LPSO-phase. At this point, the use of a powder metallurgy route or severe plastic deformation (SPD) techniques are proposed in the present study as suitable methods for refining the grain size of the alloy.

An alloy with a composition of $Mg_{98.5}Y_1Zn_{0.5}$ (at %) (with half the yttrium concentration of $Mg_{97}Y_2Zn_1$ alloy) was processed by three different techniques: ECAP, the consolidation of machining chips (CM), and powder metallurgy (PM). Therefore, the present study examines the microstructure and the mechanical strength from room temperature up to 300 °C of the $Mg_{98.5}Y_1Zn_{0.5}$ alloy processed through these alternative processes. Results are compared to those obtained in $Mg_{97}Y_2Zn_1$ alloy processed by the extrusion of the as-cast material.

2. Materials and Methods

The alloy $Mg_{98.5}Y_1Zn_{0.5}$ (at %) was obtained by melting high purity Mg and Zn as well as a Mg-22%Y (wt %) master alloy in an electric resistance furnace. The mixture was then cast in cylindrical steel molds of 42 mm in diameter. Machined chips and rapidly solidified (RS) powders were obtained from the cast bars. On one hand, the machined chips were fabricated by high-speed machining in a conventional Computer numerical controlled (CNC) lathe. RS powders were prepared by the EIGA (Electrode Induction Melting Gas Atomization) process by TLS Technik GmbH (Bitterfeld-Wolfen, Germany). Powders have a spherical morphology with a diameter less than 100 µm. Chips and powders were uniaxial compacted under 350 MPa for 2 min at room temperature (RT), providing cylindrical green compacts of 40 mm in diameter. Chip and powder compacts and cast cylinders machined up 40 mm were extruded at 350 °C using several extrusion ratios depending on the material: 4:1 and 18:1 for the cast alloy and 36:1 for the chip and powder compacts. A higher extrusion ratio is required for chips and powders to ensure their complete consolidation.

Cylinders from the extruded cast alloy bar, 70 mm long and 20 mm in diameter (extrusion ratio 4:1) were ECAP processed at 300 °C, using route B (i.e., 90° rotation of the sample between passes). A hydraulic machine with a circular cross-section die with a diameter of 20 mm and a die angle of 118° was used. Per pass, a true strain of 0.7 was produced. For ECAP process, samples were heated in the die, reaching die temperature in 5 min before pressing. A standard pressing speed of 20 mm/min was used in all cases. Following each ECAP pass, the heated split-die was opened hydraulically for rapid sample removal and then water quenched. The alloy able to be processed for up to four passes without cracking. Figure 1 shows the schematic view of the production methods for the $Mg_{98.5}Y_1Zn_{0.5}$ alloy.

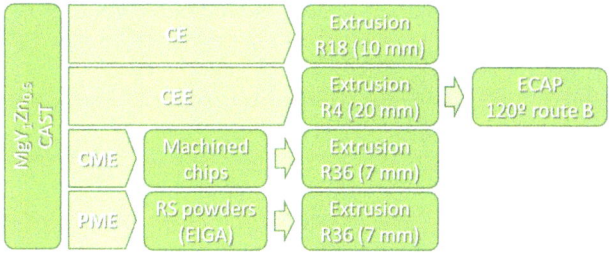

Figure 1. Schematic view of the production methods for the $Mg_{98.5}Y_1Zn_{0.5}$ alloy: CE, CME, CEE and PME.

Microstructural characterization was carried out using scanning electron microscopy (SEM) and X-ray diffraction (XRD). The microscope JEOL JSM 6500F (JEOL, Akishima, Tokio, Japan) was used in the backscattered mode. Metallographic preparation consisted of mechanical polishing and etching in a solution of 0.5 g picric acid, 5 mL acetic acid, 20 mL ethanol, 1 mL water, and 25 mL methanol. Quantitative image analysis was carried out to follow the evolution of the recrystallized fraction and grain size in the magnesium matrix after extrusion or ECAP. For the recrystallized fraction, several SEM images from areas recrystallized to different extents were measured to give a good statistical measure of this fraction. The sizes of recrystallized grains were measured, counting a minimum of 500 grains from backscattered electrons images. Statistical analyses were carried out with the software Sigma Scan Pro (Jandel Scientific, San Rafael, CA, USA), taking the grain size as the average value obtained. XRD patterns were carried out in a Siemens diffractometer D5000 using Cu-Kα radiation with a wavelength of 0.1506 nm.

Cylindrical samples machined with their long direction parallel to the extrusion/ECAP direction (head diameter of 6 mm, curvature radius of 3 mm, gauge diameter of 3 mm, and gauge length of 10 mm) were deformed in tension at a constant strain rate of 10^{-4} s^{-1} from room temperature up to 300 °C.

For simplicity, the different processing routes were designed as follows: CE for the alloy extruded from the cast bars, CEE for the ECAP processing, CME for the alloy produced using machined chips, and, finally, PME for the alloy fabricated using RS powders.

3. Results and Discussion

The micrograph of the Mg$_{98.5}$Y$_1$Zn$_{0.5}$ cast alloy (Figure 2a) revealed a two-phase dendritic microstructure consisting of magnesium dendrites with a second phase located at the interdendritic space. The interdendritic phase was characterized by a lamellar morphology typical of the LPSO-phase. The volume fraction of the LPSO-phase was 9%—much lower than volume fraction in the Mg$_{97}$Y$_2$Zn$_1$ alloy, which was between 20% and 25% [17,18]. The microstructure of the machined chips (Figure 2b) was similar to that of the cast alloy, but the LPSO-phase was broken. The dendritic microstructure of spherical RS powders was also dendritic, but much finer than that found in the as-cast ingots, as shown in Figure 2c. XRD patterns of Figure 2d confirmed that magnesium and 18R-LPSO-phase are the only phases that existed in the as-cast alloy and RS powders.

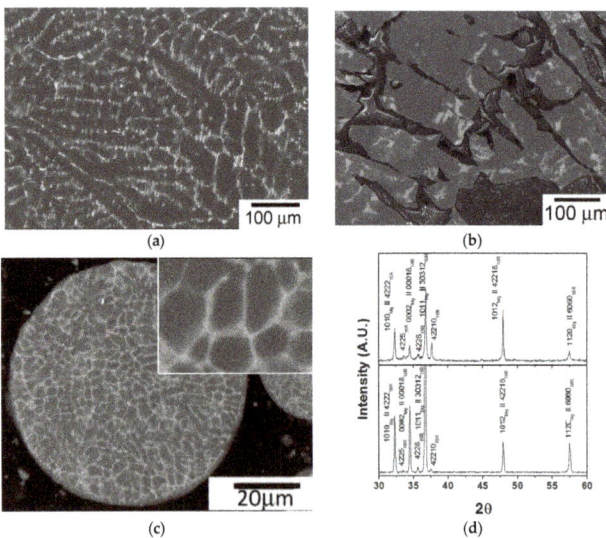

Figure 2. Microstructure of the Mg$_{98.5}$Y$_1$Zn$_{0.5}$ alloy: (**a**) as-cast alloy, (**b**) machined chips, and (**c**) rapidly solidified (RS) powders. The grey and white phases correspond to the magnesium and LPSO-phases. (**d**) Diffraction pattern of the Mg$_{98.5}$Y$_1$Zn$_{0.5}$ alloy: as-cast alloy and RS powders.

After extrusion, the LPSO-phase in the CE material appeared strongly elongated along the extrusion direction, independent of the extrusion ratio (Figure 3a). The LPSO-phase in the extruded bar produced using machined chips (the CME material) was also deformed and elongated along the extrusion direction (Figure 3b). However, at higher magnifications, it was observed that the LPSO-phase was fragmented into small particles with a diameter lower than 1 µm. The LPSO-phase in the CEE material was still elongated along the initial extrusion direction but also showed a serrated shape distribution after each pass by ECAP, as shown Figure 3c. During the extrusion process of the RS powders (PME material), the LPSO-phase appeared to be homogeneously and finely distributed in the magnesium matrix with a particle size below 500 nm (Figure 3d). These LPSO-phase particles were located mainly at grain boundaries, inhibiting the grain growth during the extrusion process.

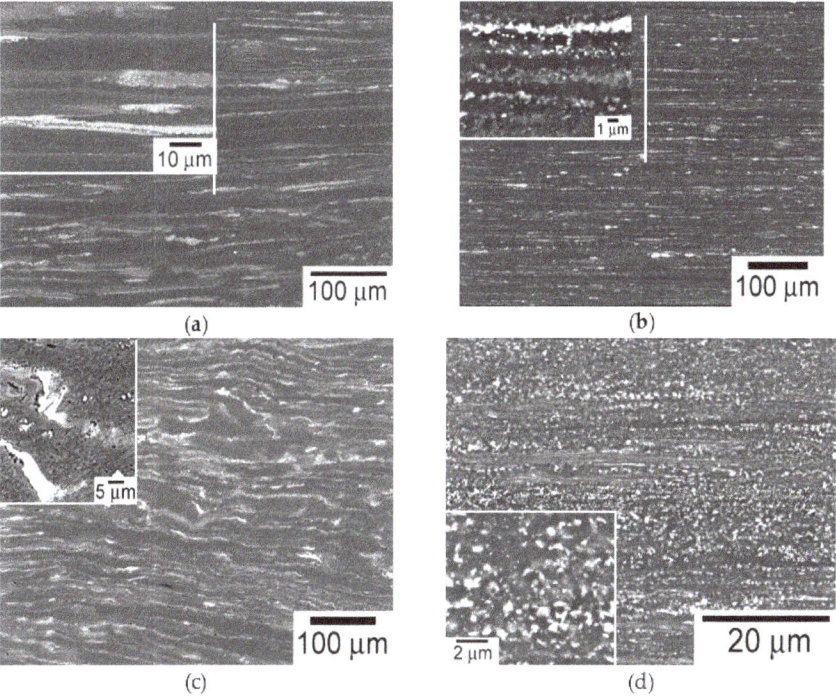

Figure 3. Microstructure of processed $Mg_{98.5}Y_1Zn_{0.5}$ alloy: (**a**) CE, (**b**) CME, (**c**) CEE, and (**d**) PME. The grey and white phases correspond to the magnesium and LPSO-phases.

The grain structure of the alloy processed by the four routes is shown in Figure 4. The grain structure depends on the processing route; it was (i) equiaxed for PME and CME alloys, and (ii) bimodal, in which fine dynamically recrystallized (DRXed) and coarse non-recrystallized grains coexist. Using image analysis techniques, the ratio of DRXed and non-DRXed regions as well as the size of DRXed grains were evaluated (see Table 1).

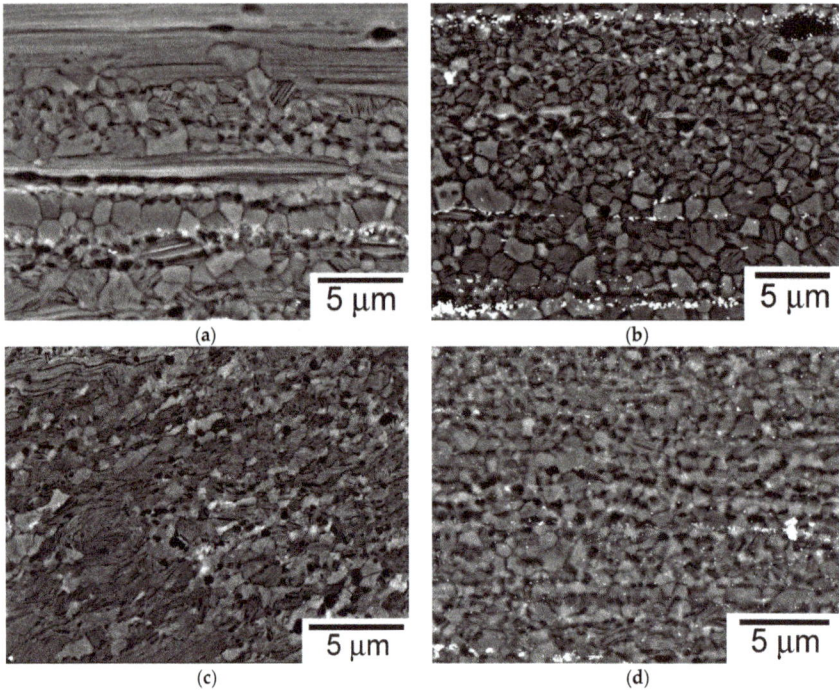

Figure 4. Grain structure of processed $Mg_{98.5}Y_1Zn_{0.5}$ alloy: (**a**) CE; (**b**) CME; (**c**) CEE; and (**d**) PME.

Table 1. Total processing strains, volume fraction of the LPSO-phase, non-recrystallized and recrystallized grains, recrystallized grain size, yield stress, $\sigma_{0.2}$, ultimate tensile strength, Ultimate tensile strength (UTS), and tensile ductility for the processed $Mg_{98.5}Y_1Zn_{0.5}$ alloy: CE, CME, CEE, and PME.

	ε_T	f_{LPSO} (%)	f_{Def} (%)	f_{DRX} (%)	D (μm)	$\sigma_{0.2}$ (MPa)	UTS (MPa)	Ductility (%)
CE	2.9	9 ± 0.7	45 ± 9	46 ± 8	1.4 ± 0.02	300	370	8
CME	3.6	9 ± 0.7	0	91	1.08 ± 0.02	313	361	13
CEE	4.2	9 ± 0.7	-	-	0.66 ± 0.02	364	383	3
PME	3.6	9 ± 0.7	0	91	0.96 ± 0.02	316	370	16

The grain size histogram of DRXed areas is showed in Figure 5. The CE material showed a bimodal grain structure with DRXed grains of about 1 μm and non-DRXed coarse grains elongated along the extrusion direction. It has been reported that DRXed grains are randomly oriented while coarse grains are oriented with the basal plane parallel to the extrusion direction [3]. The CME material was fully recrystallized with a grain size slightly lower than that of the extruded as-cast ingots (1.08 μm). The alloy processed using ECAP was not completely recrystallized. However, since both kinds of grains were intimately mixed, an accurate estimation of the volume fraction of both areas is not easily achieved without significant error. The DRXed grain size was lower than that for the CE material, 0.66 μm against 1.4 μm. Finally, the PME material was fully recrystallized with a grain size of 0.96 μm. The presence of the LPSO-phase as small particles rendered a more homogeneous distribution of this phase within the magnesium matrix, acting as pinning points which prevented further coarsening of the DRXed grains. This fact, however, did not lead to the alloy with the finest grain size, which was obtained in the alloy processed by ECAP, where the total processing strain was the highest (Table 1) and the processing temperature was the lowest.

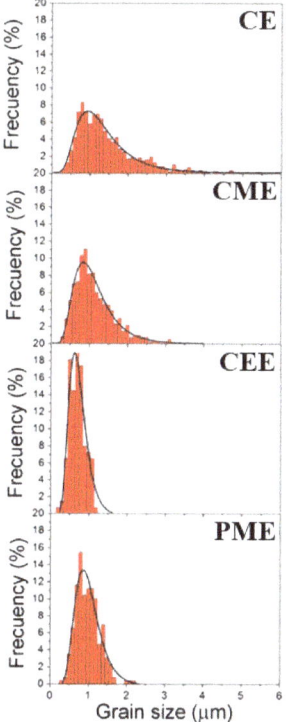

Figure 5. Grain histograms of processed $Mg_{98.5}Y_1Zn_{0.5}$ alloy: CE, CME, CEE, and PME.

Figure 6 shows the true stress-true strain curves of the $Mg_{98.5}Y_1Zn_{0.5}$ alloy at room temperature for the four processing routes. For comparison, the tensile behavior of the extruded $Mg_{97}Y_2Zn_1$ alloy is also shown [18]. It can be verified that lowering the yttrium and zinc content by the half, i.e., decreasing the volume fraction of the LPSO-phase in the alloys processed by the extrusion of as-cast ingots, induced a significant decrease of about 150 MPa, although the elongation to failure was enhanced.

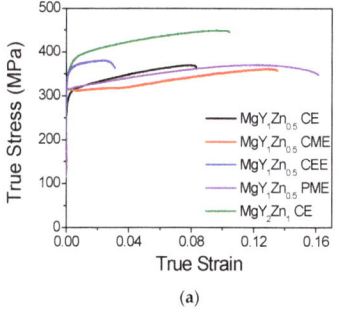
(a)

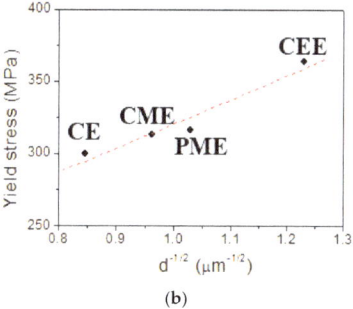
(b)

Figure 6. (a) True stress-true strain curves of processed $Mg_{98.5}Y_1Zn_{0.5}$ alloy: CE, CME CEE, and PME. For comparison, the curve of the extruded $Mg_{97}Y_2Zn_1$ alloy obtained from the as-cast alloy is included; (b) Evolution of the yield stress as a function of the inverse square root of the grain size for the alloy processed by the four routes: CE, CME, CEE, and PME.

The curves of materials processed by machined chips or RS powders, MCE and PME in Figure 6, revealed similar yield stresses, 313 and 316 MPa, respectively. These values are slightly higher than that of the conventional extruded material, 310 MPa, but the ultimate tensile stresses (UTS) are lower. Yamasaki et al. [3] proposed that coarse grains contribute to reinforce the alloy since neither basal slip nor tensile twinning can be easily activated in these grains. Garces et al. [19,20] confirmed this assumption using diffraction methods during in situ tension and compression tests. Moreover, they proved that the beginning of plasticity (yield stress) is always controlled by the activation of the basal slip system in DRXed grains. Therefore, the higher yield stress of CME and PME materials is due to their finer grain size. In the case of the CE alloy, coarse grains areas (non-DRXed grains) contribute to reinforce the material during tensile tests, reaching a yield stress almost comparable to that of CME and PME alloys. It is important to mention that the material produced using RS powders showed the higher elongation to failure. The use of the PM route in magnesium alloys produced extruded bars with low texture and, therefore, favored the ductility [21]. The $Mg_{98.5}Y_1Zn_{0.5}$ processed by ECAP, the CEE material, attained the same yield stress as the $Mg_{97}Y_2Zn_1$ alloy, although its ductility was highly reduced. This fact confirms again that the beginning of plasticity in this kind of alloys with bimodal grain structure is controlled by the plasticity of DRXed grains, so the grain refinement of these grains is the best strategy to improve the mechanical strength in these alloys. Figure 6b shows the variation of the yield stress with the reciprocal of the square root of the grain size ($d^{-1/2}$). The yield stress increased as the grain size decreased following a Hall-Petch relationship with a slope of 170 MPa $\mu m^{-1/2}$, similar to that obtained by Hagihara et al. [4] in $Mg_{97}Y_2Zn_1$ alloy.

Figure 7 shows the true stress-true strain curves of the $Mg_{98.5}Y_1Zn_{0.5}$ alloy for the four processing routes from RT to 300 °C at 10^{-4} s^{-1}. For all processing routes, two ranges of behavior were distinguished: (1) the interval from RT up to 200 °C with high strength, high work hardening, and low ductility, and (2) the interval from 200 to 300 °C in which a pronounced decrease in strength was noticed but elongation to failure was substantially improved. The evolution of the yield stress is shown in Figure 8a from RT to 300 °C for the four processing routes. In all materials there was a gradual decrease in strength as the temperature increased, although the strength remained above 200 MPa up to 200 °C, and even up to 250 °C in the case of the CEE material. The alloy processed by extrusion followed by ECAP exhibited the highest yield stress over the entire temperature range. The mechanical behavior was determined mainly by two parameters: (i) the equiaxed or bimodal structure of the magnesium matrix, and (ii) the size and morphology of the LPSO-phase. The use of chips, powders resulted in fine-grained equiaxed magnesium grains. In addition, the alternative processing routes induced changes in the LPSO-phase. In the case of the CME material, the LPSO-phase was broken during the machining stage. In addition, the large strain accumulated during machining induced subsequent fracture during the extrusion process. In the CEE material, however, most of the LPSO-phase particles were not broken during ECAP. Instead, they were severely kinked. The distribution and alignment of the LPSO-phase was similar to that of the CE material, i.e., the LPSO-phase was arranged in long strings of particles aligned along the extrusion direction. However, the linearity of such arrangements was lower, especially in the case of the CEE material, compared to the CE alloy (as can be seen in Figure 1). A major difference of PME alloy was that the fine dendritic structure of the RS powders results in the fracture of LPSO-phase particles located at the interdendritic regions, leading to very homogenous distribution in the magnesium matrix. At room temperature and 100 °C, it seemed that the effect of grain size prevailed over the reinforcement due to the LPSO-phase. The reinforcing effect of LPSO-phase particles, larger than 0.5–1 μm in size, on the strength of magnesium alloys through the load transfer mechanism is well reported [20,22], due to its higher Young Modulus compared to the magnesium phase [22,23]. Thus, up to 100 °C, grain size is the main parameter controlling the yield stress of the different alloys. As the volume fraction of the LPSO-phase is essentially the same in all alloys and no large differences in the grain size of DRXed regions, changes in the reinforcing effect due to load transfer from the magnesium matrix to the LPSO-phase should be small, as a result of differences in the shape and size of the LPSO-phase particles.

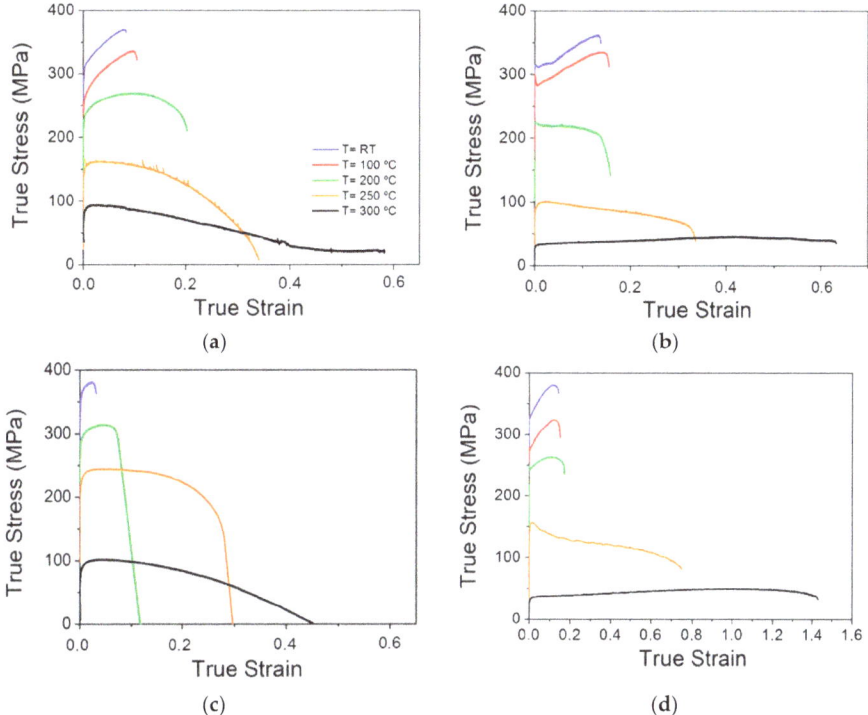

Figure 7. True stress-true strain curves in the temperature range of 25–300 °C. (**a**) CE, (**b**) CME, (**c**) CEE, and (**d**) PME.

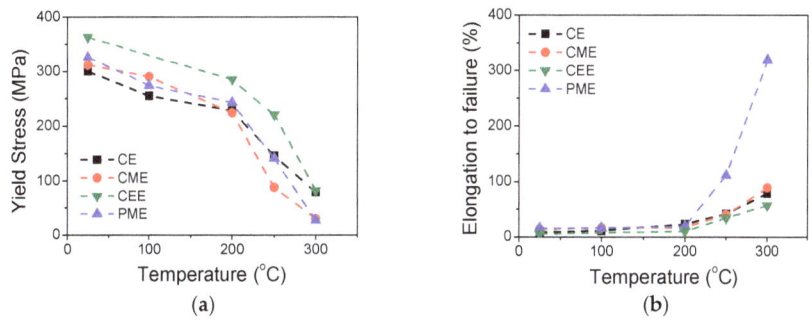

Figure 8. Evolution of the (**a**) yield stress and (**b**) elongation to failure as a function of the temperature for the four processing routes.

Above 200 °C, both the specific characteristics of the magnesium matrix and the LPSO-phase particles in each alloy determined their mechanical behavior. Thus, the alloys with a fully recrystallized microstructure (CME and PME) became much softer than alloys with a bimodal microstructure (CE and CEE). Such different behavior was associated with some contribution of the grain boundary sliding (GBS) mechanism during the plastic deformation of the alloy. It is well reported that magnesium alloys containing the LPSO-phase exhibit superplasticity, even in alloys reinforced with volume fractions of coarse second phases exceeding 50% [6,24,25]. Consequently, GBS can proceed not only between magnesium grains but also in LPSO-phase/Mg interfaces. In alloys with grain sizes ranging between 2 and 5 µm and coarse second phases, superplastic behaviors with high elongations have been attained at temperatures above 300 °C because the accommodation of stresses generated during GBS must be achieved by cracking the LPSO-phase particles, which subsequently become homogeneously redistributed in the magnesium matrix. In addition, the existence of coarse non-DRXed regions precludes the activity of GBS there, reducing the superplastic deformation capabilities of the material. Superplastic behavior could take place once these coarse non-DRXed grains become recrystallized, usually at temperatures above 300 °C. Given the special features required for GBS in alloys containing the LPSO-phase, only the PME alloy exhibited superplastic behavior at 300 °C, 317% as elongation to failure (see Figure 8b), because this alloy was the only one combining the grain refinement of the magnesium matrix and the homogeneous distribution of fine second phases.

4. Conclusions

Severe deformation processing has been used to decrease the yttrium content of Mg-Y-Zn alloys containing LPSO phases without losing their mechanical strength at room temperature. In this study, three alternative routes were explored to maximize the mechanical strength of the $Mg_{98.5}Y_1Zn_{0.5}$ alloy: the extrusion of machined chips and RS powders obtained from the cast alloy, and ECAP processing of the extruded bar. The microstructure of the extruded bars obtained from the machined chips and RS powders was fully recrystallized with a grain size near 1 µm, slightly improving the yield stress in comparison to the extruded alloy. On the other hand, the use of ECAP processing highly refined the grain size of the alloy up to 660 nm in such a way that its yield stress became similar to that of the extruded alloy with twice the yttrium content. Superplasticity could be achieved at 300 °C for the alloy prepared by a powder metallurgy route due to the grain size of the alloy, and the homogeneous distribution permitted the operation of grain boundary sliding.

Author Contributions: Each author equally contributed to the development of the paper. G.G., S.C., and R.B. carried out the processing of the materials and their microstructural characterization. P.P. contributed to the microstructural and mechanical characterization of the materials. P.A. contributed to the paper writing and interpretation.

Funding: This research was funded by the Spanish Ministry of Economy and Competitiveness under project number MAT2016-78850-R and the Czech Grant Agency under grant Nr. 16-12075S

Acknowledgments: Authors would like to acknowledge the expert support of Miguel Acedo and Edurne Laurin for assistance with extrusion processing and metallographic preparation, respectively.

Conflicts of Interest: The authors declare no conflict of interest.

References

1. Inoue, A.; Kawamura, Y.; Matsushita, M.; Hayashi, K.; Koike, J. Novel hexagonal structure and ultrahigh strength of magnesium solid solution in the Mg–Zn–Y system. *J. Mater. Res.* **2001**, *16*, 1894–1900. [CrossRef]
2. Kawamura, Y.; Kasahara, T.; Izumi, S.; Yamasaki, M. Elevated temperature $Mg_{97}Y_2Cu_1$ alloy with long period ordered structure. *Scr. Mater.* **2006**, *55*, 453–456. [CrossRef]
3. Yamasaki, M.; Hashimoto, K.; Hagihara, K.; Kawamura, Y. Effect of multimodal microstructure evolution on mechanical properties of Mg–Zn–Y extruded alloy. *Acta Mater.* **2011**, *59*, 3646–3658. [CrossRef]
4. Hagihara, K.; Kinoshita, A.; Sugino, Y.; Yamasaki, M.; Kawamura, Y.; Yasuda, H.Y.; Umakoshi, Y. Effect of long period stacking ordered phase on mechanical properties of $Mg_{97}Zn_1Y_2$ extruded alloy. *Acta Mater.* **2010**, *58*, 6282–6293. [CrossRef]
5. Garces, G.; Oñorbe, E.; Dobes, F.; Pérez, P.; Antoranz, J.M.; Adeva, P. Effect of microstructure on creep behaviour of cast $Mg_{97}Y_2Zn_1$ (at %) alloy. *Mater. Sci. Eng. A* **2012**, *539*, 48–55. [CrossRef]
6. Oñorbe, E.; Garces, G.; Dobes, F.; Pérez, P.; Adeva, P. High-Temperature Mechanical Behavior of Extruded Mg-Y-Zn Alloy Containing LPSO Phases. *Metall. Mater. Trans. A* **2013**, *44*, 2869–2883. [CrossRef]
7. Luo, Z.P.; Zhang, S.Q. High-resolution electron microscopy on the X-$Mg_{12}ZnY$ phase in high strength Mg-Zn-Zr-Y magnesium alloy. *J. Mater. Sci. Lett.* **2000**, *19*, 813–815. [CrossRef]

8. Abe, E.; Kawamura, Y.; Hayashi, K.; Inoue, A. Long-period ordered structure in a high-strength nanocrystalline Mg-1 at% Zn-2 at% Y alloy studied by atomic-resolution Z-contrast STEM. *Acta Mater.* **2002**, *50*, 3845–3857. [CrossRef]
9. Garces, G.; Pérez, P.; Gonzalez, S.; Adeva, P. Development of long-period ordered structures during crystallization of amorphous $Mg_{80}TM_{10}Y_{10}$. *Int. J. Mater. Res.* **2006**, *4*, 404–408.
10. Zhu, Y.M.; Morton, A.J.; Nie, J.F. The 18R and 14H long-period stacking ordered structures in Mg–Y–Zn alloys. *Acta Mater.* **2010**, *58*, 2936–2947. [CrossRef]
11. Egusa, D.; Abe, E. The structure of long period stacking/order Mg–Zn–RE phases with extended non-stoichiometry ranges. *Acta Mater.* **2012**, *60*, 166–178. [CrossRef]
12. Garces, G.; Maeso, M.; Todd, I.; Pérez, P.; Adeva, P. Deformation behaviour in rapidly solidified $Mg_{97}Y_2Zn_1$ (at %) alloy. *J. Alloys Compd.* **2007**, *432*, L10–L14. [CrossRef]
13. Garces, G.; Muñoz-Morris, M.A.; Morris, D.G.; Perez, P.; Adeva, P. Optimization of strength by microstructural refinement of MgY_2Zn_1 alloy during extrusion and ECAP processing. *Mater. Sci. Eng. A* **2014**, *614*, 96–105. [CrossRef]
14. Garcés, G.; Muñoz-Morris, M.A.; Morris, D.G.; Pérez, P.; Adeva, P. Maintaining high strength at high temperature in a Mg–Y–Zn–Gd alloy by heat treatments and severe deformation processing. *Metall. Mater. Trans. A* **2015**, *46*, 5644–5655. [CrossRef]
15. China's Rare Earth Industry and Export Regime: Economic and Trade Implications for the United States. Available online: http://www.fas.org/sgp/crs/row/R42510.pdf (accessed on 12 May 2016).
16. Japanese Motor Manufacturers Looking for Alternatives to Rare-Earth Metals. Available online: http://www.greencarcongress.com/2012/01/motors-20120101.html (accessed on 12 May 2016).
17. Oñorbe, E.; Garces, G.; Pérez, P.; Adeva, P. Effect of the LPSO volume fraction on the microstructure and mechanical properties of Mg–Y_{2X}–Zn_X alloys. *J. Mater. Sci.* **2012**, *47*, 1085–1093. [CrossRef]
18. Garces, G.; Perez, P.; Cabeza, S.; Lin, H.K.; Kim, S.; Gan, W.; Adeva, P. Reverse tension/compression asymmetry of a Mg–Y–Zn alloys containing LPSO phases. *Mater. Sci. Eng. A* **2015**, *647*, 287–293. [CrossRef]
19. Garces, G.; Morris, D.G.; Muñoz-Morris, M.A.; Perez, P.; Tolnai, D.; Mendis, C.; Stark, A.; Lim, H.K.; Kim, S.; Shell, N.; et al. Plasticity analysis by synchrotron radiation in a $Mg_{97}Y_2Zn_1$ alloy with bimodal grain structure and containing LPSO phase. *Acta Mater.* **2015**, *94*, 78–86. [CrossRef]
20. Garces, G.; Pérez, P.; Cabeza, S.; Kabra, S.; Gan, W.; Adeva, P. Effect of extrusion temperature on the plastic deformation of an Mg-Y-Zn alloy containing LPSO phase using in-situ neutron diffraction. *Metall. Mater. Trans A* **2017**, *48*, 5332–5343. [CrossRef]
21. Cabeza, S.; Garces, G.; Pérez, P.; Adeva, P. Properties of WZ21 (%wt) alloy processed by a powder metallurgy route. *J. Mech. Behav. Biomed. Mater.* **2015**, *46*, 115–126. [CrossRef] [PubMed]
22. Oñorbe, E.; Garcés, G.; Pérez, P.; Cabezas, S.; Klaus, M.; Genzel, C.; Frutos, E.; Adeva, P. The evolution of internal strain in Mg–Y–Zn alloys with a long period stacking ordered structure. *Scr. Mater.* **2011**, *65*, 719–722.
23. Tane, M.; Nagai, Y.; Kimizuka, H.; Hagihara, K.; Kawamura, Y. Elastic properties of an Mg-Zn-Y alloy single crystal with a long-period stacking-ordered structure. *Acta Mater.* **2013**, *61*, 6338–6351. [CrossRef]
24. Pérez, P.; Eddahbi, M.; González, S.; Garces, G.; Adeva, P. Refinement of the microstructure during superplastic deformation of extruded $Mg_{94}Ni_3Y_{1.5}CeMM_{1.5}$ alloy. *Scr. Mater.* **2011**, *64*, 33–36. [CrossRef]
25. González, S.; Pérez, P.; Garces, G.; Adeva, P. Influence of the processing route on the mechanical properties at high temperature of Mg-Ni-Y-RE alloys containing LPSO-phase. *Mater. Sci. Eng. A* **2016**, *673*, 266–279. [CrossRef]

© 2018 by the authors. Licensee MDPI, Basel, Switzerland. This article is an open access article distributed under the terms and conditions of the Creative Commons Attribution (CC BY) license (http://creativecommons.org/licenses/by/4.0/).

MDPI
St. Alban-Anlage 66
4052 Basel
Switzerland
Tel. +41 61 683 77 34
Fax +41 61 302 89 18
www.mdpi.com

Materials Editorial Office
E-mail: materials@mdpi.com
www.mdpi.com/journal/materials

www.ingramcontent.com/pod-product-compliance
Lightning Source LLC
LaVergne TN
LVHW071958080526
838202LV00064B/6782